Thomas H. Elsom

Spokane's First Telephone Installer

Dedicated to

Thomas H. Elsom

His biography and the early history of the telephone in the Spokane area would not have been possible without his diary record and hundreds of early photographs that he took and developed.

What would he say about today's latest telephone technology? With the launching of communication satellites we now have hand-held global satellite phones and paging networks covering the entire earth. Furthermore, we communicate via e-mail and the Internet. Yet these advances have built upon the hard work and ingenuity of telephone pioneers such as Thomas H. Elsom.

Thomas H. Elsom

Spokane's First Telephone Installer

Dean Ladd

Tornado Creek Publications
Spokane, Washington 2000

Copyright © 2000 by Dean Ladd
and/or
the Spokane Corral of the Westerners
and/or
Tornado Creek Publications

All rights reserved, including the rights to translate or reproduce this work or parts thereof in any form or by media, without the prior written permission from the author or publisher. In the event of the death of the author or publishers all rights to this publication shall revert to the Spokane Corral of the Westerners.

Published in 2000
Printed in the United States of America
by
Walsworth Publishing Company
Marceline, Missouri

Library of Congress Control Number: 00-090726
ISBN (hard cover): 0-9652219-5-4

Cover photo: Thomas Elsom and the Inland Northwest's first telephone installation crew. Standing from left: Sultzman, Blanton and Kollo. Seated are Black and Thomas Elsom. This photo was taken in 1888, while the crew was installing the new telephone system in Wallace, Idaho.

Tornado Creek Publications
Tony and Suzanne Bamonte
P.O. Box 8625
Spokane, Washington 99203-8625

Acknowledgements

Two of Thomas Elsom's daughters, Gertrude Ladd and Floral Stephenson, preserved considerable information about their father's telephone pioneer record. Gertrude typed Thomas's 1931 handwritten recollections. These records, along with his many diaries, notes and hundreds of photographs provide a great wealth of historical information.

Grandson Larry Elsom spent many years pulling all the pieces together and developed an extensive computer filing system to readily access the information stored in many file cabinets and shelves. The final step was to make it available for the present document. This is the culmination of years of intermittent effort by various descendants of Thomas Elsom. In 1999 the Elsom Collection was donated to the Eastern Washington State Historical Society.

Unless indicated otherwise, all photographs are from the Thomas Elsom collection.

Dean Ladd

Compiler Dean Ladd, another grandson of Thomas Elsom, was born December 8, 1920 in Spokane and graduated from Bemis Elementary and Rogers High School. He enlisted in the Marine Corps Organized Reserve in April 1939 and completed one year at Washington State College before being called to active duty in November 1940.

Vera and Dean Ladd

He received a field commission and saw extensive WWII combat in the Pacific with the Second Marine Division at Guadalcanal, Tarawa, Saipan and Tinian as recalled in his book, *Faithful Warriors*. He was wounded three times, the most seriously during the landing at Tarawa. After five years he returned to get his mechanical engineering degree and was an engineer three years at Kaiser Aluminum, six years at North American Aviation and 23 years at Lockheed on aerospace programs. He remained active in the Marine Reserve for a total of 30 years, retiring as a lieutenant colonel.

One year after college graduation, he married Vera Michel from high school days. They have three daughters and 17 grandchildren.

This publication was
revised, edited and published by
Tony Bamonte, Suzanne Bamonte
and Laura Arksey

Indexing by
Ed Weilep

Table of Contents

CHRONOLOGY   8

CHAPTER 1: Entrepreneurial Beginnings   11

CHAPTER 2: Thomas Elsom: Inland Northwest Telephone Pioneer   32

CHAPTER 3: Inland Northwest Telephone Expansion in Thomas Elsom's Words   73

CHAPTER 4: Retirement and Family Photo Album   112

INDEX   123

CHRONOLOGY

1865 Thomas Elsom born August 11, in Carlton, New York.
1876 Alexander Graham Bell demonstrated the first telephone. Custer's defeat at Little Big Horn happened the same day.
1877 Charles B. Hopkins founded the first newspaper in the Inland Empire at Colfax and started planning use of the telephone for his newspaper business. There was a Nez Perce Indian uprising that same year.
1878 Military telegraph line installed from Fort Missoula to Fort Walla Walla due to Indian uprising.
1879 *Spokan Times*, Spokane's first newspaper, started by Francis Cook.
1880 Fort Spokane established by Gen. O.O. Howard. Spokane Falls population was about 350.
1881 Northern Pacific Railroad reached Spokane Falls.
1882 Cheney Academy (now EWU) opened.
1883 Gold rush to the Coeur d'Alene mountains begins.
Spokane Falls Review newspaper started as a weekly May 15 and became a daily the next year.
Thomas Elsom started working part time on the railroad and on the family farm in Dakota Territory.
1884 C. B. Hopkins installed the first telephone toll line in the state between Colfax and Almota on the Snake River for his Colfax newspaper. Telephone service first established in Seattle and Tacoma.
1886 Thomas Elsom arrived in Spokane Falls October 5 and on November 11 was hired by C. B. Hopkins. In late December, he installed Spokane's first telephone set in the grocery store owned by Walker L. Bean at Howard and Riverside. Forty years later, he installed the first dial phone in the same location.
1887 First toll telephone line was installed on west side of state between Tacoma and Puyallup.
Gonzaga University founded.
Hopkins bought entire army telegraph line from Fort Missoula

to Fort Spokane and Fort Walla Walla for $20.00 Started as C.B. Hopkins Telephone System installing long distance telephone line. Elsom began working on the lines in the Coeur d'Alene mining areas. For $565, William S. Norman in turn bought from Hopkins a switchboard located in Spokane Falls and the army telegraph line, which extended to the Montana border. He leased 30 telephone boxes from the Bell Company.

1888 First telephone lead installed between Spokane Falls and Seattle. Many mining camps now reached by telephone. 160 telephone sets and 600 miles of line served from Spokane Falls.

1889 Washington became a state.
Downtown Spokane destroyed by fire August 4. Elsom was working in Idaho at time of fire.

1890 Industrial Exposition held in Spokane Falls. Rapid rebuilding of the city with financing from insurance claims and other massive entrepreneurial investing. Elsom became foreman of telephone construction and a year later chief foreman for the Inland Telephone and Telegraph Company.

1891 C. B. Hopkins Telephone Company became the Pacific States Telephone Company in June.
Elsom married Nell Pratt.

1892 Telephone service established between Spokane and Portland and shortly extended to San Francisco.

1893 Panic of 1893. Hopkins transferred telephone company headquarters from Colfax to Spokane. Shortly afterward he organized under license from American Bell Company. He became general manager, operating telephone business throughout the Inland Territory. This region included that portion of Washington east of the Cascades, eastern Oregon and part of Idaho.

1898 First automobiles appear in Spokane.
1899 Fort George Wright opens in Spokane.
1900 Telephone lines extended from Waterville to Lake Chelan.
1901 Telephone lines extended from Wenatchee to Cle Elum and connect to a line built in 1900 from Yakima to Seattle.

1905 Elsom was district plant superintendent until 1911 when he left the company for two years.
1915 Telephone lines completed from New York to San Francisco. The Pacific Company acquired property of the Home Telephone and Telegraph Company in Spokane. The name was changed to Pacific Telephone.
Elsom became a telephone right-of-way agent until retirement.
Division Street Bridge collapsed on December 18
1924 Party at Davenport Hotel in honor of Elsom's 35th year with the telephone company.
1926 50th anniversary of the telephone invention.
1930 Elsom retired from telephone company September 1 after 41 years of service.
1953 Elsom died December 15 at age 88.

Telephone operators in Spokane, circa 1920.
(Photo courtesy Eastern Washington State Historical Society, L95-13.1)

Chapter 1
Entrepreneurial Beginnings

The telephone is 100 years younger than the birth of our nation. On June 25, 1876, Alexander Graham Bell had demonstrated his telephone at a Centennial Exposition in Philadelphia, the same day as the Custer massacre at the Little Big Horn. Fortunately for Bell, the Emperor of Brazil saw the demonstration and took a liking to Bell. The emperor convinced the panel of judges to take closer notice. (One of them, Lord Kelvin, later became a famous physicist.) In 1872 Bell had started producing the first crude drawings and experiments for a "harmonic telegraph" that would distinguish between musical notes to transmit over a wire. Then he experimented with speech and hearing with equipment using the ear of a dead man. This led to his discovery of the principle of the telephone in 1874. A few months later, with the aid of his machinist assistant, Thomas Watson, he was in a race with another experimenter, Elisha Gray, who later became the cofounder of the Western Electric Company.

In March of 1875 the inventor of the magnetic telegraph, Joseph Henry, advised Bell to drop all else and press on with the telephone rather than trying to improve telegraphy. His experiments led to an accidental breakthrough several months later while adjusting transmitting reeds to vibrate at tuned pitches.

The first electric speaking telephone consisted of a wooden frame on which was mounted a harmonic receiver with one end of its steel-reed armature touching a tightly stretched membrane of parchment. Because of its shape it became known in telephone history as the "gallows" telephone.

On February 14, 1876, Bell filed a patent application for his device. A few hours later Elisha Gray filed a "caveat," warning other inventors of a speaking telephone. Bell's application had a single paragraph on the variable-resistance principle written into the margin of a page. The patent became the subject of thousands of pages of testimony from challengers in hundreds of unsuccessful suits.

Alexander Graham Bell in 1876, the year he invented the telephone.

Thomas A. Watson, Bell's very able machinist assistant in 1874 (left) and in San Francisco in 1915 at transcontinental telephone line opening (right).

*(Photos on page 12-15 from **Telephone, The First Hundred Years** by John Brooks. N.Y.: Harper and Row, 1975.)*

EVERY
MAN, WOMAN AND CHILD
SHOULD CAREFULLY EXAMINE THE WORKINGS OF
PROF. BELL'S
SPEAKING AND SINGING TELEPHONE

In its practical work of conveying
INSTANTANEOUS COMMUNICATION BY DIRECT SOUND
Giving the tones of the voice so that the person speaking can be recognized by the sound at the other end of the line.
Having secured a large number of Prof. A.G. Bell's TELEPHONES will give an EXHIBITION at the OLD JOHN STREET M.E, CHURCH
44 & 46 JOHN ST, NY
Where all visitors desiring can make for themselves a practical investigation of the TELEPHONE, by asking questions, hearing the answers to the questions and listening to the singing conveyed through the telephone from the other end of the line.
On Tuesday and Wednesday Afternoons
November 20th & 21st, 1877
From 111/2 A.M. until 7 P.M.

The liquid telephone over which the historic words, "Mr. Watson, please come here, I want you," were spoken on March 10, 1876. The exhibition notice above was for various sets in November 1877.

Elisha Gray, the Chicago inventor. His claims to having invented the telephone came closest of many such claims to overthrowing Bell's.

Francis Blake, Jr. The Blake transmitter, invented in 1878, employed carbon and greatly improved telephone service.

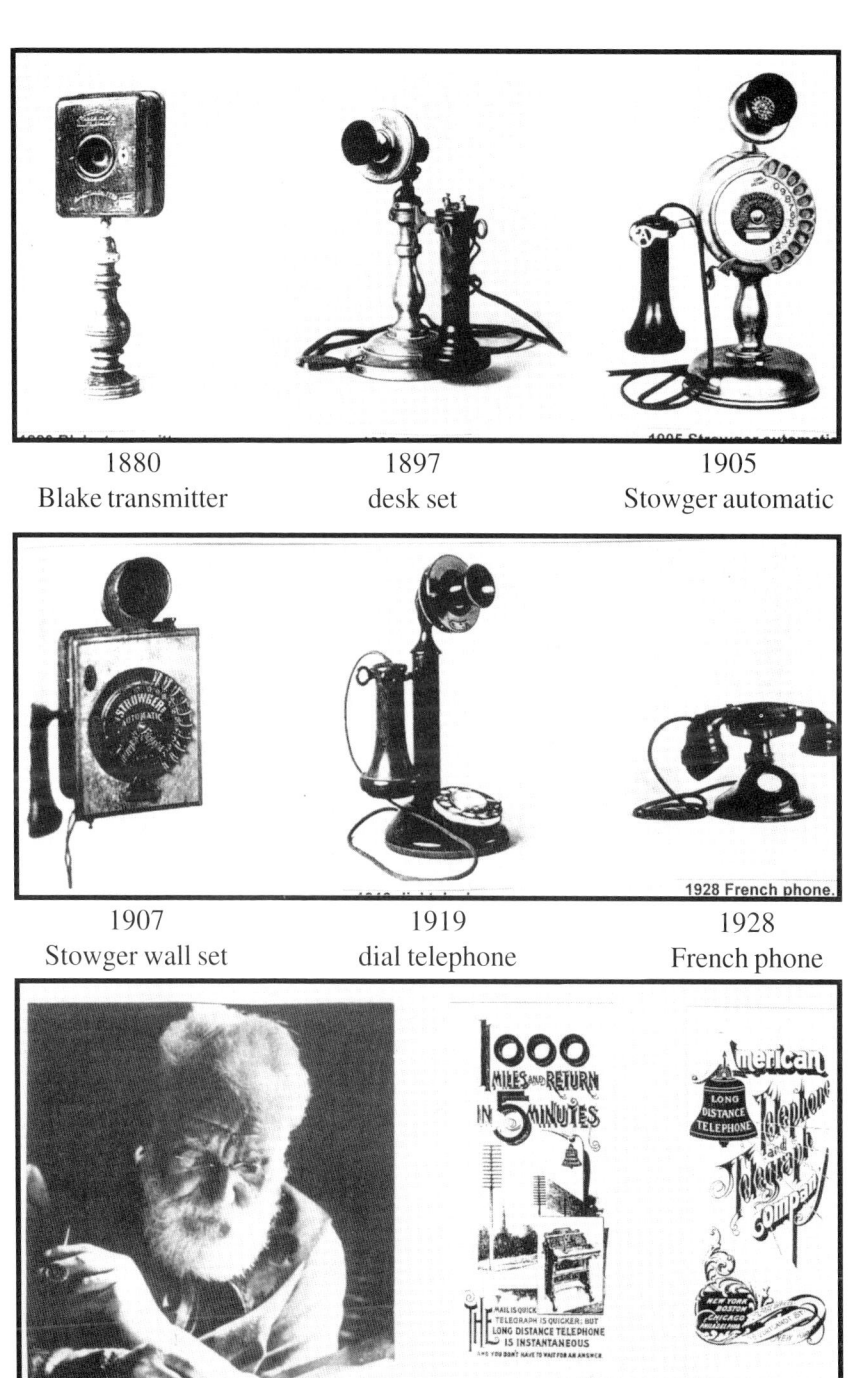

Evolution of the earliest telephone sets. Bottom left: Bell in later life.

The Bell Telephone Company was created a year after the demonstrations. For the next two years, it had trouble competing with Western Union Telegraph in conducting even the simplest conversation. Much litigation also continued between the Bell Telephone Company and competitors until the turn of the century, especially over the patent for the "Berliner" transmitter (microphone).

In 1881 two-wire circuits started replacing one-wire circuits insulated from the ground. The first underground telephone cables were laid in 1882 but did not become widespread until a decade later.

Charles B. Hopkins

A few years later, the telephone came to the Inland Northwest. The first community to use it, in a limited way, was Walla Walla, where the telephone connected an express office and a post office and between a lumber mill and the end of a flume

Colfax entrepreneur Charles B. Hopkins was among the first to realize the potential for the newspaper business of this new communication device. In 1877 he and Lucien E. Kellogg had founded at Colfax the *Palouse Gazette*, probably the first newspaper in the entire Inland Empire north of the Snake River. Colfax was then an agricultural and trading metropolis of about 400. The first newspaper in the smaller community of Spokane Falls was the *Spokan Times* (no "e" on Spokan) with the first issue in April 1879.

Hopkins took the lead in establishing the telephone business in the Inland Empire when there was no railroad or telegraph. Connection with the outer world was by a three-day steamboat trip down the Snake and Columbia rivers to Portland. The boat was met by stagecoach and freight wagon at Almota sixteen miles south of Colfax. The Northern Pacific Railroad was still under construction from St. Paul.

The outbreak of the Nez Perce war was causing turmoil throughout the area, and the United States Signal Service began constructing an emergency telegraph line from Fort Missoula to the various army outposts in northern Idaho and eastern Washington. However, the

Indian trouble ended before the line was completed. It had been routed past Spokane Falls through Colfax and across the Snake River. The government operated it for only a short time and then abandoned it.

In looking back on this period, Hopkins said, in an address to telephone employees in San Francisco on September 21, 1915.

> In 1883 I first heard about the telephone through Dr. Joseph Jorgensen, recently arrived from Virginia, who had served in Congress, and failing of re-election, had come west with a commission as receiver of the U.S. Land Office and an ambition to become Senator from the about-to-be-admitted State of Washington. The doctor brought a set of telephones west with him. In furtherance of his ambition to become Senator, he cultivated my friendship, doubtless figuring upon the assistance my newspaper might render. He told me about the telephones and presented them to me.
>
> I secured government consent to install them upon the old military telegraph line; one at Colfax and one at Almota, thus establishing the first long-distance telephone connection in the Inland Empire and practically the first in Oregon, Washington and Idaho.
>
> Quite a little patronage developed and with it my ambition to extend the service. Not so much because of the money there was in it, but because of the advantage it would afford me in gathering news and skinning the rival newspapers recently established in neighboring communities.
>
> About this time a section of the military telegraph line was condemned and ordered sold by the Quartermaster at Fort Walla Walla at 9 o'clock on the morning of a certain day to the highest bidder for cash. I endeavored to borrow some cash from the local Shylocks, but without success. They regarded the telephone as a plaything and me as crazy in wanting to put money into it. Anyway, I went to Walla Walla on credit for stage fare and hotel bills and showed up at the Quartermaster's office a little before 9 o'clock with a lonesome twenty dollars in my pocket and the hope that no competing bidders would be there, and strange as it may seem, this pipe dream came true. There were no other bidders and 100 miles of fine telegraph line was knocked down to me for twenty dollars, and a bill of sale was written on the margin of one of the posters advertising the auction.

Cartoon of Charles B. Hopkins in 1890. He was the founder of the first newspaper, the *Palouse Gazette,* and established the first telephone business in the Inland Empire. The source of this cartoon was Thomas Elsom's album. Under the cartoon, Elsom had written the following caption: "The Boss the way to Olympia to mend fences."

My sensations were those of exultation and joy mingled with regret that I had bid so much. Nevertheless, there was another thrill coming, for hardly had I pocketed the bill of sale than the door was opened and in strutted a Walla Walla barber, commissioned by a relative, a merchant and a banker at Lewiston, Idaho, to bid $250, and he blusteringly demanded that the sale to me should be set aside and his tender accepted.

The Quartermaster looked troubled and my heart went into my boots, but only for a moment. The officer had the courage to stand by what was right and just, and informed the barber he might do as he pleased about protesting to the commanding officer. Thus I broke into the telephone game for a $20 gold piece.

After considerable negotiating and the reluctant surrender of the telephones, an extra-territorial concession was secured from the American Bell Telephone Company and the balance of the military telephones also became my property. Then black and white signs reading "C.B. Hopkins Telephone System" graced the front of many long-distance offices operated on a commission of 10% to the agent, which also compensated him for keeping a certain length of line in repair, the honor and free talks also helping some.

Extensions to towns not along the route of the military line were somewhat inferior in construction. The advice of San Francisco experts was solicited and accepted. The specifications were poles 12 or 15 feet long set 200 to 220 feet apart; porcelain insulators, and if these ran short, the neck of a bottle was a good substitute.

Hopkins initially planned to operate lines only in connection with his paper, the *Palouse Gazette.* But the system grew to such proportions and became so profitable that he took in a partner, who had exclusive charge of the paper while Hopkins devoted his entire time to the telephone. He extended the line to Spokane in 1886 and a telephone exchange was established in the Hyde Building on the southeast corner of Riverside and Wall.

The following quote, which summarizes Hopkins's role with the early telephone system, was taken from a historical account written by

Gertrude Elsom Ladd, Thomas Elsom's daughter, from information recorded by Elsom.

> In 1887 and 1888 Mr. Hopkins bought a section of the old government telegraph line and established a connection by telephone between Coeur d'Alene and Wardner, Idaho, and also between Spokane and Davenport, and extended the Colfax-Almota line to Fort Walla Walla.
>
> After the big fire of August 1889, the native pine poles were replaced with 65-foot cedar poles shipped in from Idaho. By 1900 there were so many wires that it was necessary to place them underground. The company then constructed its own building at N. 117 Wall Street.
>
> In about 1878, the United States Government had built a single #9 (1/8-inch in diameter) iron-wire line from Fort Missoula, Montana, via Wardner and Fort Sherman (now Coeur d'Alene), Idaho, to Spokane, Colfax, Almota, Lewiston, etc. . . . In the development of this new country, the town of Colfax naturally became the principal distributing point for a large section called the Palouse Country, for which the steamboat landing at Almota on the Snake River was the shipping point. This resulted in considerable business between Almota and Colfax. About this time, Mr. C.B. Hopkins became identified with the development of the town of Colfax and established a newspaper, etc., and at that point, recognizing the advantage of having a ready means of communication between Colfax and its shipping point at Almota, he secured telephone instruments. He arranged for use of the government telegraph wire and in 1883 or 1884, he installed one at Colfax and one at Almota, thus establishing a toll telephone line about eighteen miles long.
>
> This was the inception of the telephone system of Eastern Washington, and Mr. Hopkins . . . continued to extend it by the purchase of the government telegraph line and construction of extensions to the different towns as they came into existence and developed business to justify the establishment of telephone connections. By 1886, the system had reached the principal towns of the Palouse Country in the vicinity of Colfax and extended to Spokane and Walla Walla.
>
> About this time the silver-lead mines of the Coeur d'Alene district, center

ing around Wardner, Idaho, came into prominence and the prairie country west of Spokane known as the "Big Bend" country began to develop, as did the timber and mining industries north of Spokane. This brought Spokane into prominence as the natural commercial center of a large territory, which included north Idaho, part of Montana and British Columbia.

Mr. Hopkins acquired the government telegraph line from Spokane to Wardner, Idaho, and from Spokane west, passing through the Big Bend country to Fort Spokane on the Columbia River near the mouth of the Spokane River. Thus the center of the telephone system of Eastern Washington was transferred from Colfax to Spokane, which became the headquarters of the Inland Telephone and Telegraph Company. This company was shortly afterwards organized under license from the American Bell Telephone Company with C.B. Hopkins as general manager . . .

William S. "Billy" Norman

Apparently Hopkins decided to focus on the phone connections already established in his area. This line extended to the Snake River for rapid communication with Portland, rather than from the Spokane Falls area toward the Montana line. The latter portion he offered to William S. "Billy" Norman, a visionary businessman who would also soon become an important part of young Thomas Elsom's life work.

An informative biography about Billy Norman, written by historian-author Jerome Peltier, was published in 1988 in *The Pacific Northwesterner* (Volume 33, No. 2). Excerpts from that article, entitled "Billy Norman – Frontier Capitalist," are included, as follows:

> The way most people picture an empire builder is far different from the appearance of the short, soft-spoken, dapper gentleman that I once knew named William S. Norman. It is easy to understand why most of his friends called him "Billy," for he was easily approached and generous with his time and ideas. His smile was infectious. He was born in Cheltenham, Gloucestershire, England, January 8, 1858. His father ran a printing and lithographing shop and was also the editor and publisher of two newspapers in Cheltenham. Billy worked for his father before coming to Washington Territory in February 1883.

William S. Norman, owner of the Inland Telephone and Telegraph Company in 1889. Elsom worked for him as the first telephones were installed in Spokane and the Coeur d'Alene mining camps. (*Photo from "Spokane Falls and Its Exposition."*)

On arrival, Norman settled on land at Dragoon Creek, north of present day Spokane, while he worked on farms on Moran Prairie. He soon left to embark on a more exciting project. The Canadian Pacific Railroad was completing construction of the western part of its transcontinental line. Tracks were being laid from the first crossing of the upper Columbia River on the east to Revelstoke, B.C., situated at the second crossing on the west, where a bridge had to be built.

H.H. McCartney and Company held the contract for supplying all of the grain, hay and foodstuffs for the construction gangs and horses. The supplies came mainly from Washington Territory and Oregon, and Joseph and Marcus Oppenheimer, Spokane financiers, had the contract for getting them to Revelstoke. Transporting them presented a formidable problem, as there were no roads north of Northport, W.T. The Columbia River ran by Little Dalles, ten miles east of Northport. The Oppenheimers elected to build a warehouse at Little Dalles and a riverboat to carry the supplies up the river. They hired Joe Vogel to supervise the job, and Vogel hired Norman as timekeeper. When Vogel learned that Norman could transcribe dictation and write letters, he made him his secretary and raised his salary from twenty-five dollars to sixty dollars a month.

Building a lake steamer in the middle of a wilderness was a difficult task. To get lumber for the boat, a sawmill had to be freighted in and assembled. Norman wrote me in a letter, "All of the lumber employed in the building of the boat, except for the ribs, was cut out and dressed at a sawmill plant established for that purpose on Onion Creek a mile and a half above Little Dalles . . . " The ribs had to be brought in from Portland, Oregon.

Obtaining marine engines was even more of a problem . . . [At this point in the article, Peltier further details the construction of the Oppenheimers' steamer, *Kootenai*, and their shipping business. The boat was commissioned in late April 1885. Billy Norman became the purser and made countless trips carrying supplies up the Columbia River to the Canadian Pacific Railroad construction crews. As an aside, the *Kootenai* was later sold to Spokane railroad entrepreneur D.C. Corbin and was used for transportation on the Columbia to the Kootenai River junction.]

With the end of the shipping season, Billy Norman returned to Spokane and became a stenographer at the Spokane County Court House, then at the county seat in Cheney. He had learned stenography as a reporter while working on his father's newspapers in Cheltenham. Spokane was about to enter a period of rapid economic expansion fueled by numerous discoveries of mineral wealth nearby and the coming of the Northern Pacific Railway. Sagebrush-covered land was being developed into farms and a rapidly growing Spokane became the trade center for the region. It was at this time that Billy was presented with a number of opportunities that led to his meteoric rise in the world of business.

Early in 1886, Norman formed a real estate partnership with A. A. Newberry who represented the interest of the Northern Pacific Railway. Norman suddenly found himself in the company of the movers and shakers of Spokane. He noted that "A. A. Newberry had much to do with fashioning the railroad situation in the Inland Empire (a region comprised of eastern Washington, northern Idaho and Montana west of the Rockies). He was a great friend of Corbin and it was through his (Newberry's) efforts that the Spokane and Northern railroad was built by Corbin." Norman became assistant secretary to Newberry, A. M. Cannon and Paul F. Moore, all pioneer bankers, during the organization of the Spokane and Northern.

Billy learned lessons in high finance from these men and had ideas of his own that he wanted to explore. Charles Hopkins, owner of a newspaper and a telephone company in Colfax, Washington, gave Billy his first big break. For undisclosed reasons, Hopkins offered to sell the telephone company to Norman. The offer was a surprise, but Norman seized the opportunity. His letter to me of May 10, 1949, says, "My memory is pretty clear still as to the happenings of those early days, especially the successful start of the W. S. Norman Telephone System and its ramifications of long-distance lines throughout the Inland Empire. The original investment as a starter was $565 which was given to Charles B. Hopkins to release the equipment – a switchboard for fifty lines, thirty telephone boxes under rental lease from the Bell Company and sundry diverse equipment required in the installation. In selling this equipment, Charlie Hopkins threw in part of the system from Spokane County through to the Montana line via the Wallace and Kellogg (Idaho) area."

Norman could not provide the entire cost of the system himself, but brought in his friends who made up the difference. His partners were S.Z. Mitchell who later became one of the executives of the General Electric Industry of America, now known as the General Electric Company; Burt Nichols, who later built the Nichols Block on Riverside Avenue in downtown Spokane; and Lieutenant Fred Spalding. Mitchell and Spalding were both engineers with the Edison Electric Illuminating Company, one of the early electric utility companies in Spokane. Eventually S. Z. Mitchell became president of the Electric Bond and Share Company, a large electric utility holding company.

As soon as the telephone equipment reached Spokane Falls, it was placed in the front room of the Norman and Newberry real estate office in the Hyde Building. When the company began business, it had thirty-five telephones available and only fourteen subscribers. Billy was not dismayed. During the next few months, he bought out his partners.

N. W. Durham, Spokane historian, states that Norman influenced the U.S. War Department to condemn its telegraph line between Coeur d'Alene City and Fort Sherman to Spokane.

During an interview I had with Norman, he gave me a different version: "I paid twenty-five dollars for the telephone line from Spokane to Fort Sherman. This line ran from Walla Walla to Missoula, following the old Mullan Road. It was government owned and operated at this time. Towns through which they (the lines) ran were the eventual owners of the lines, with the stipulation that the lines be maintained in good order.

The lines had been down three days when I went to Coeur d'Alene to see Colonel Carlin . . . We were introduced and I said, "Colonel, I would like to buy the telephone line." He said he'd write and ask permission to sell. He got permission (to do so) and advertised it in the newspaper as required, and I eventually bought it, after signing an agreement to the effect that I would keep it in good running condition.

Apparently Norman misspoke himself in describing the telegraph line as a telephone line, as he later changed this statement.

Norman related in a letter: "A Montana syndicate, headed by D. C. Corbin, had secured a contract to move 50,000 tons or more of ore from the recently discovered Bunker Hill mine. The tonnage was handled across Lake Coeur d'Alene in lake boats and then transported over a spur track from Fort Sherman to Hauser Junction (Idaho) to East Helena (Montana) for smelting. Afterward, as other mines were developed, the Corbin road gridironed the Coeur d'Alenes as far as Burke and Mullan on the east. At that time and before the Corbin road was built, I had purchased the U.S. Military telegraph line, from Coeur d'Alene City through the Fourth of July Canyon to Cataldo Mission and thence to the Montana Line. In 1886, it was repaired and equipped with telephones. The lines were extended as Corbin extended his rail lines. My telephone company gave free use of service on all of its lines in exchange for right-of-way and the transportation rights for a rail velocipede for repair purposes.

After completion of the first National Bank Building in 1887, on the corner of Riverside and Howard Streets, Billy rented the third floor and basement for the telephone office. By this time, 250 subscribers were using the company, and connections had been made with telephone exchanges in Wardner, Murray, Mullan and Burke, Idaho, in the Coeur d'Alene mining district. This was only eleven years after Alexander Graham Bell had invented the telephone.

In 1884, the Edison Electric Illuminating Company had built a hydroelectric power station on the site of the present Upper Falls plant of the Washington Water Power Company. Billy Norman bought stock in the company in 1887 and became active in its management. At the time he joined the company, the power plant was small and the water supply uneven in quantity, so he immediately tried to improve the situation. In 1888 he got an option on all of the waterpower rights on the Spokane River west of Post Street and east of Monroe by paying $485,000 for the C and C Flouring Mill and the Post Mill, which held the rights. The Washington Water Power Company was formed as the entity to buy the rights and to purchase the Edison Company and its power station. To raise capital for further expansion, the company enlisted the assistance of merchant bankers in Brooklyn, New York. The company then built a dam across the Spokane River and a power station at the foot of Monroe Street, completing the planned project.

The joy of accomplishment was overcome by the great fire of 1889, which destroyed the Spokane telephone system. Fortunately, the small Edison power station was not burned. Unfortunately, the fire destroyed all of the Washington Water Power poles and distribution system.

The fire was a great economic leveler and resulted in consolidation of some local businesses. Norman and Hopkins sold one-half of their interest in Inland Telephone and Telegraph Company to Sunset Telephone and Telegraph Company, which at that time represented a large part of the interests of the Bell Telephone Company in the Pacific states.

The street railway system needed help at that time of crisis too. The city was fragmented and each part had its separate system. J. J. Browne and A.M. Cannon owned a horse-car system and G. B. Dennis was the principal owner of a partially completed electric street railway known as the Ross Park Line. The Spokane Cable Railway was also partially completed. All of these lines were acquired by the Washington Water Power Company, which proceeded to unify and electrify the entire system. The power company also monopolized all of the city's lighting system.

The panic of 1893 stopped any plans for further business expansion. Most businesses strove merely to survive. Washington Water Power was able to stay afloat, but in so doing, many of its promoters lost heavily. Norman was one of these and he had to sever his connections with the company in 1898 because he could no longer pay his required assessments. [Peltier's article continues to discuss Norman's business ventures following this period, after which he was no longer involved with the telephone business. He and his brother Ben Norman purchased the Spokane Hotel, which they remodeled. It became Spokane's finest hotel until the Davenport was built. The Normans went on to develop a chain of hotels.]

The entrepreneurship of Hopkins and Norman has to be seen in the light of the bustling growth of Spokane in the 1880s and 1890s. Fortunes of men like Glover, Browne and Cannon were made and lost. Cannon died alone and bankrupt in his New York hotel room. Developments in land, agriculture, timber, railroads and mining were stunning, and the telephone played a significant part. Of course, there were ups and downs. Spokane's Great Fire of 1889 and the Panic of

1893 were major setbacks to the city, the region and individuals. The fire caused Spokane to rebuild the city with large, distinguished buildings of stone and brick. The Panic of 1893 resulted in the failure of banks and land speculation, the unemployment of thousands and such movements as Coxey's Army, an 1894 march of desperate men to Washington, D.C. But after each setback, there was recovery, and the importance of the telephone increased.

An event that reflected the basic entrepreneurial optimism of the period was the Spokane Falls 1890 Industrial Exposition, an impressive venture so soon after the fire. Thomas Elsom, like many other in Spokane, saved the large souvenir publication produced for the Exposition. Of particular interest is the large Exposition Building, a mammoth wooden structure, 200 by 300 feet and three-stories high. It was located on the corner of Sprague and Hatch Road, across the street from the present Becker Buick. The wooden building had a very short life, being destroyed by fire in 1893, the same year as the Panic. One of the items in the Exposition publication is a map showing the routing of railroads, horse-car lines, electric motor lines and a cable car line.

Part of the recovery of Spokane was the beginning of Fort George Wright, and telephone entrepreneur William S. Norman was involved. In 1894 he was one of a group of businessmen who learned of plans for an army post in the Northwest. This group was successful in bringing the fort to Spokane by donating 1,000 acres costing $40,000. With the help of Spokane women, who organized fund-raising events, they were able to secure their city as the location of this important post, which became Fort George Wright.

With his telephone development and chain of hotels, William Norman continued as prominent empire builder, involved in the Republican Party and many social organizations. He died in 1954, at the age of 96, in the same house, located at 644 West Seventh, where he had lived since the fire of 1889. The entrepreneurial career of Charles Hopkins, though distinguished, was much shorter than that of Norman, as he died in 1920.

Thomas Elsom was not himself an entrepreneur. He was, however, a valuable employee of Hopkins and Norman. With his technical ability, apt leadership of the crewmen, and innovative developments, he enabled these entrepreneurs to successfully develop the telephone system of Spokane and the Inland Northwest.

Over the years, as the telephone system was being installed and developed in the Inland Northwest, Thomas Elsom photographed and processed hundreds of images chronicling much of the region's early history. All the photographs beyond this point, unless otherwise noted, were either taken by Elsom or were part of his personal collection.

Sprague Avenue, looking east from Howard Street, circa 1890, showing telephone and electric trolley lines. The Oakes Cafe is at the left.

First and Post in 1888

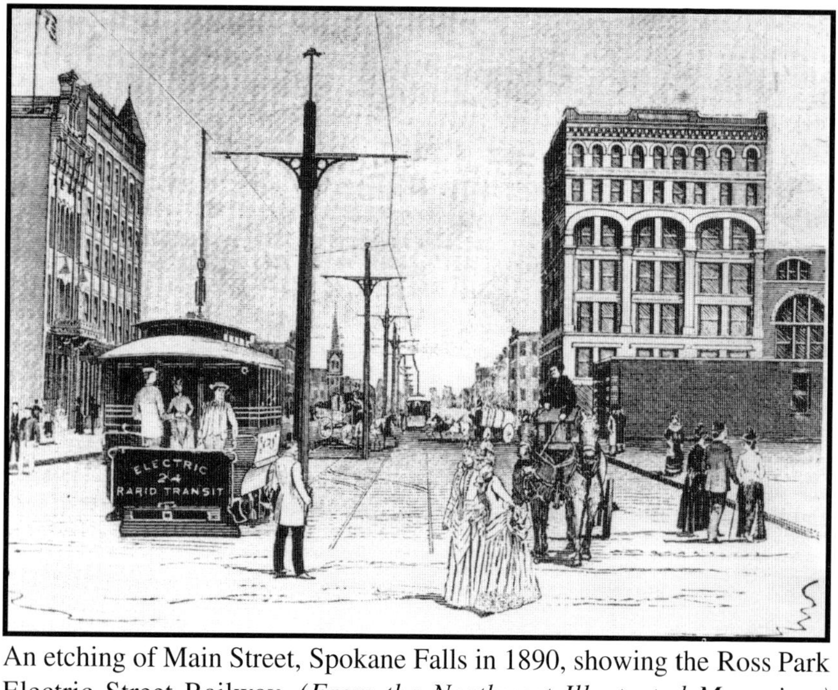

An etching of Main Street, Spokane Falls in 1890, showing the Ross Park Electric Street Railway. *(From the Northwest Illustrated Magazine.)*

Thomas Elsom in 1888.

Chapter 2
Thomas Elsom:
Inland Northwest Telephone Pioneer

Thomas Henry Elsom entered this world in Carlton, Orleans County, New York, on August 11, 1865. The son of Joseph and Jane (Harmer) Elsom, he was the third of nine children. His brothers and sisters, in birth order (surviving to adulthood), were Anna L., Charles W., Evert J., Wilson J. and Mary F..

Thomas was born nine months after his father had been discharged from three-and-a-half-years Civil War service with the 8th New York Volunteer Cavalry. His parents had married while still teenagers in August 1860, one year before Joseph enlisted.

Joseph was discharged December 15, 1864, miraculously unwounded after fighting in 49 of the 54 engagements in which his regiment had participated. Those included such major battles as Harpers Ferry, Fredricksburg, Gettysburg, Second Battle of Bull Run, Antietam, Beverly Ford and Sherman's march through Georgia to the sea.

A few years after Thomas's birth, the growing young family moved to a farm at nearby Oak Orchard. Then in 1881, they moved to the Midwest and homesteaded two miles south of Northville, South Dakota (then Dakota Territory). Joseph became the first postmaster at Northville and served a term in the state legislature.

An important technological milestone had occurred five years before that would profoundly affect young Thomas's later career. Alexander Graham Bell had successfully demonstrated his telephone in 1876.

January 1, 1883, at age 17, Thomas started keeping a diary. Coincidentally, this was the same year the telephone first appeared in Washington Territory at Seattle. Then a year later, it appeared in Tacoma, followed a short time later in Walla Walla, Port Townsend, Spokane Falls, Colfax, Yakima, Ellensburg and Olympia, in that order.

Above: Thomas's mother and father, Joseph and Jane (Harmer) Elsom, in 1860 as newlyweds. Below left: Thomas Elsom's father, Joseph, in 1862, as a 22-year-old Civil War cavalryman. Below right: Joe Elsom holding flag.

The Elsom Diaries

Thomas Elsom faithfully kept up a diary until two years before his death. The earliest entries provide a window into his emerging manhood on the homestead in South Dakota with his family and working for the telephone company in Spokane. It is likely his earlier employment on the telegraph line may have prepared him somewhat for telephone work.

Elsom's diaries are a resource gold mine and are perhaps even more valuable than a personal interview, because there is no memory loss. They detail his daily activities supervising early telephone systems construction. His responsibilities covered the entire Northwest from the Canadian border into Oregon and from the Montana state line to the Cascades. Some selected diary entries are as follows:

1883 Diary Entries

January
2nd Working on railroad making shims. **5th** Inspecting track. **9th** Freeze ears. **15th** Main rail line blocked so no mail from the East. **22nd** Driving spikes. Find broken bolts.

February
8th Quit railroad. Work on farm. **12th** Back working on railroad.

March
1st Across Minnesota border shoveling snow off tracks. **29th** Dig a well 17 ft. deep. Get good water.

April
4th Get kicked in the leg by horse. **16th** Plowing all day.

May
12th Finish sowing oats. **24th** Working this forenoon for grandmother.

1884 Diary Entries

June
29 Father gets hurt with his bull.

July
1st Bull gets after me and rolls me in the grass. **2nd** Bull knocks Mr. Gomer out of tree. **7th** Uncle shot bull with rifle this evening.

August
4th Begin harvesting day and night. **9th** Finish cutting. **11th** Begin stacking. My 19th birthday. **23rd** Work for father in warehouse.

December
23rd Get feet frozen. Get big pair of boots and overshoes for $5.00.

1885 Diary Entries
January
1st Working on railroad. At Highmore, Dakota, repairing telegraph wire. Go on hand car 3 miles and repair brake in 48. Go to Huron. Go 4 miles east on Velocipede and fix brake. Weather cold. Has been 10 to 45 below zero for three weeks. **2nd** Go to Hurley on freight this morning. Fix brake six miles south and walk back to town. Wait all night for train. It doesn't come. Freeze my fingers. Get dinner at Canerota for 25 cents. Bed and board at Hurley 75 cents **13th** [He was starting to seriously consider going west. He entered details about Montana from the *World Encyclopedia* into the end of his 1885 diary.] **16th** Go out on freight train. Install 20 rods of new [telegraph] wire four miles north of Hitchcock. Nail up two brackets, flag passenger [train] and go to Northville. Cold. **19th** In Huron adding water to batteries. Jars fine. 20 below zero.

February
6th Go to DeSmet this morning. Put on one bracket three miles west. Go to Huron on No. 2. [He then left this job and received pay owed him, immediately returning home to Northville to work on the family farm and odd jobs while getting more education.]

December
2nd Get some blank books to study bookkeeping.

At the end of his diary, he wrote what it took to become a civil engineer as follows: The education of those who would rise to prominence in the profession of Civil Engineering must embrace a fair knowledge of mathematics, the sciences, natural philosophy, mechanics, hydraulics and optics. They should acquire knowledge of the principles of projection and should aim at being a good draughtsman – rapid and accurate.

Elsom Arrives in Spokane Falls

Probably at about the time he arrived in Spokane in 1886, he jotted the notation "2 Kings 8:2" at the top of the last page of his diary. That Bible verse tells about a woman who followed Elisha's advice to take her household and sojourn in the land of the Philistines for seven years to escape famine in the land. It certainly must have had some significance for Elsom at the time or he would not have written this only reference to the Bible in his diary. Perhaps it was merely confirmation to him that he should go west for awhile, because times were hard in South Dakota.

1886 Diary Entries
October
1st George Crawford and self leave for Washington Territory with $60.00. **3rd** Fargo. Mislaid the passes . . . **6th** Arrived here [Spokane Falls] yesterday. The boys are looking for a boat to go down the river. Am staying in the tent and writing letters. This is a fine town but not much farming country to it. **9th** Cutting logs by mill for Mr. Cook [possibly Francis Cook]. Cut 11 logs. **13th** Cut 32 logs and milk a stray cow. **14th** Come near skipping the job but got to stick or starve. **17th** Go to work in sawmill. **24th** George Crawford leaves for Idaho to look for land, Now living in cabin with Chip and Hank. **31st** Thinking of going [further] west. Working at sawmill for $1.00 per cord.

November
10th Rent two rooms for $9.00/month. Get a job with the telephone company for $2.50/day. [This is the first diary entry relating to his telephone work in Spokane.] **12th** Taking down old [telegraph] wire. **13th** Putting up new [telephone] wire and insulators. **15th** There are six men here for every job of common labor. **18th** Get paid by Mr. Hopkins. Get letter from father. **25th** They show me about the

telephone machine and set me to work connecting the cable. **26th** Run cable into a house on Riverside. **30th** Working for the phone company in morning and electric light company in afternoon. Climbing 50-ft poles and putting up arc lights.

December

5th Buy a violin for $10.00. [Equal to more than a month's rent.] **10th** Working for the electric company. Raise the lamp at the Cannon Block. **20th** Telephone company wants me to go to Coeur d'Alene, Idaho. [He had installed the first phone in Spokane Falls at Walker Bean's grocery store just prior to this.] **21st** The boys go to Montana and want me to go. **25th** Directions for adjusting the "Berliner" transmitter [for which he had written eight pages of details at the end of his 1885 diary]. Visit the old camp on the bluff. Get a very good [Christmas] dinner. **29th** Get through working for the electric company. **30th** Go out about twenty miles to east on horseback and find downed phone poles.

The First Telephone in Spokane Falls

In 1886 there was rapid growth in Spokane. The first branch railway, the Spokane & Palouse, was built. Earlier discoveries of vast silver and lead deposits in the Coeur d'Alene region led to the construction of a Northern Pacific branch line from Hauser, Idaho, to Coeur d'Alene Lake. Spokane's first street railway was built, and both Spokane Electric Light & Power Company and Sacred Heart Hospital were organized.

The following is from an account of Elsom's career with the telephone company published shortly after his retirement in 1930. It appeared in the *Pacific Telephone Magazine* of June 1931, and was entitled, "Little Stories About Some of Us."

> 21-year-old Thomas Elsom arrived by train in Spokane Falls on October 6, 1886. [Elsom's diary entry of October 6 stated he "arrived yesterday" – the 5th.] He had been interested in becoming a locomotive engineer and had started as an engine wiper in the frontier roundhouse of the Chicago & Northwestern Railway at Huron, Dakota Territory.

His mother was concerned about the hazards of railroading so he started training to be a telegraph operator. This soon bored him. One day a man came along with spurs and climbers and said, "If you want a real job, come with me." Thomas went. He was still in the telegraph service, but out in the open where there were poles to climb, troubles to fix and adventure to find. He installed and repaired telegraph lines for nearly two years in Huron before deciding to seek adventure in the wild West and possibly get a job with the Northern Pacific Railroad.

He arrived with $60 and started living with two or three other young men in a tent. His first work was cutting logs for $1.00 a cord. He said work was hard to find. There were six men to every job. He mentioned working for Signors and for Francis Cook. He said he didn't like this type of work but had to work or starve. He mentioned in his diary, "baking some bread that was hard as stone and graham flour does not agree with us very well."

He told of working for the Electric Light Company "guying poles and trimming lamps." Then he was sent to the Coeur d'Alenes by the U.S. Government to locate poles that were down. He repaired the telegraph line at $2.50 a day or $5.00 a day for himself and a horse. One time he rode thirty miles by horseback and found three miles of wire down.

Then, going up the street, he saw a man trimming poles. Upon inquiring, he learned that they were for a new telephone system being installed. He further learned that C. B. Hopkins was the owner of the entire system in Eastern Washington. On November 11, 1886 he found Hopkins and asked for a job. Hopkin's response was, "If you can climb a pole, go to work."

He immediately started work as a lineman, shaving and framing poles for starting the Spokane exchange. During that winter he helped place a line of poles on Riverside Avenue from Post to Bernard Street and on Howard from Front [now Spokane Falls Boulevard] to the Northern Pacific Railroad. He later reminisced, "I couldn't fall off the poles if I tried. They were pine and oozed so much pitch."

When the poles were in place and the pioneer Spokane subscribers were connected, Thomas became the whole gang – lineman, installer, repairman, troubleshooter – everything about the plant. "And at night," he

Detail from a larger sketch of Spokane Falls in 1884 as recreated in 1944 by artist William Donohue. *(Courtesy Spokane Public Library.)*

Corner of Howard and Riverside looking west in 1889 before the fire. Thomas Elsom installed the first telephone in Spokane Falls in the grocery store at the left in December 1886. The First National Bank Building on the right was built in 1887. The first Spokane telephone exchange moved from the Hyde Building and rented the third floor and basement of the bank.

recalled, "I slept on a cot in the office in order to answer the night calls."

Thus destiny guided the ambitions of this young man away from locomotives to become a pioneer in the picturesque development of the telephone in the Pacific Northwest.

In late December 1886, Elsom installed the first telephone set in Spokane for the Inland Telephone and Telegraph Company in the grocery store at Howard and Riverside of real estate broker Walker L. Bean. Forty years later, he installed the first dial telephone at the same location.

He described the first subscriber set as three wooden boxes, screwed one above the other to the wall on a backboard about eight to ten inches wide and three feet long. The top box was a magneto generator with a crank that was turned to call the central office and had a hook for the receiver. Below this was a smaller box containing a transmitter. Below that, the third box contained the battery.

On Christmas Day, 1886, he made a diary entry about a many-page technical description of how to adjust a transmitter. He was constantly reading, even on Christmas Day, to improve on his job.

The central office was on the ground floor of the Hyde Building on Riverside and Wall (then Mill Street). The exchange consisted of one 50-wire switchboard. The first operator, Lucius G. Nash, later became an attorney.

A line of 25-foot native pine poles was constructed along the north side of Riverside Avenue between Post and Stevens Streets and along Howard from Front (now Spokane Falls Boulevard) to First Avenue to Browne's Addition. The poles were cut nearby. The lines were all single wire rather than two as later installed.

1887 Diary Entries
January
3rd Take charge of telephone exchange today. Busy all day. **12th** Putting new telephones together.

Telephone inside the Spokane telephone office in 1887.

Sample pages from Thomas Elsom's personal diaries.

NOVEMBER, SUNDAY 13. 1887.
go to Gardner with Ren fix his Instrument and do some woork on line get back late rain

NOVEMBER, MEMORANDA. 1887.

MONDAY 14.
go to wolf Lodge find line down in 2 places get there late get letter from sis J. Coleman and word wright,

THURSDAY 17.
at mission A. M. get letter from Mother go to camp Hardie pay P. $5— get the machine owe him $10 snow soon as learn the B.

TUESDAY 15.
go out to Blue Creek Canyon fix up line and meet Mr. Sipson about 2 miles above the canyon rain, get home late tonight

FRIDAY 18.
at mission all day fine. Henry packed up to move the store Mr. Philsbery here looking over things

WEDNESDAY 16.
at mission all day fine get letter from Cora H,

SATURDAY 19.
at Mission A. M. go to camp Hardie this P. M. get some paper print a couple of pictures.

September, Sunday 11. 1887.

out fishing all
day rainy get 53
good ones
write to Meggie

Monday 12.

Receiving Hay for
Ham & Roy this A. M.
go out fishing with
Henry this P. M. —
get a sponge for Boat
$1.00

Tuesday 13.

at mission
go with Henry
in the boat to find
the shingle in lango
to fix deck on Schid
and get wet. fine.
get a letter from J. L. S.

Wednesday 14.

get new fishing tackle
Pole   4.00
line   .75   catch mess for
reel   1.25   supper fine
leader   .60
flies   .—
         3.25
hooks

September, Memoranda. 1887.

Thursday 15.

go to Wolf Lodge and
repair A brake in
wire fine day

Friday 16.

at mission fish a little
get letter from
Sis & Fred A. fine

Saturday 17.

tighten up the wire
at mission and go
fishing catch about
20 fine write to J. L. S.

February
28th Tell Nichols [his boss] that I need to have a raise in wage.

March
2nd Get one month's pay – $40.00. **23rd** Get a new suit of clothes for $15.00. Get a raise in wages.

April
5th City elections today. Attend wake, speeches and bonfire. Get hit in leg with dynamite cartridge. **8th** Funeral today. They were drowned in the steamboat accident. **21st** Man ground to death by train today. **27th** Spokane Falls has 7,000 inhabitants. **30th** Drew my pay, sold my spurs, paid board bill.

Thomas then became discouraged and decided to go back home to South Dakota. He returned to Spokane Falls two months later. His diary often subtly revealed such brief ups and downs.

May
1st Leave Spokane Falls on train for East at 10:45 A.M. Arrive at Huron, Montana, at 3:40 P.M. Huron a hell of a place. Nothing but saloons and breweries. Noxon nothing but station and wood shed. Arrive at Thompson Falls at 6:15. Built near river and lots of Indians. **4th** Back in Northville. Go to work in warehouse unloading feed and flour.

June
23rd Go and tell Mrs. Coleman and Maggie good-bye. Leave Northville by train for Spokane at 4 P.M. Fare is $35.00. **27th** Arrive in Spokane about 1 A.M. Checking phone lines on horseback.

He immediately went back to work on the same day of arrival. He received a salary increase from $40.00 to $75.00 per month. The salary information was obtained from a document found in his billfold after his death.

July
1st At Coeur d'Alene. **3rd** In Spokane helping move telephone

exchange. **13th** Arrive at Wardner to help repair phone lines.

July
29th Help load train car with wild horse. Come near getting neck broken.

August
3rd Go up to summit above Kingston to repair line. **10th** Put in telephone poles at Wardner and move telephone office. **23rd** Thrown by horse. Hurt leg and shoulder.

December
22nd Studying photography. [This is his first entry about photography.] **28th** Make my first negative. [His photos used in this document are among the almost 2,000 remaining in the collection.]

Expansion of the Inland Northwest's Telephone System

There were 35 original subscribers [accounts of this number vary widely, but this is from Elsom's account] connected to the Spokane switchboard. Expansion of the system was slow. Many prospects for telephone service said they were not interested until a telephone was installed in the Northern Pacific Depot, which would enable them to call for the arrival time of the trains.

When William Norman bought the telephone equipment from Charles Hopkins in 1887, it also had included the remnants of the United States military telegraph line that followed the old Mullan Road from Walla Walla to Missoula. Initially, the purchase did not include the section between Spokane Falls and Fort Sherman at Coeur d'Alene. This section was still owned by the government, but Norman shortly also bought that link for only $25.00, with the stipulation that the lines would be maintained in good order.

The First National Bank Building was completed in 1887 on the corner of Howard and Riverside. The Spokane telephone exchange rented the third floor and basement. By now there were 250 customers and connections were made to exchanges in the Coeur d'Alene mining

Wardner, Idaho, circa 1900.

The Wardner-Wardner Junction stagecoach, the Black Mariah, in 1887.

district at Wardner, Murray, Mullan, Wallace and Burke. The first calls there began July 18, 1887. By then, subscribers numbered 950 and paid $1.00 per minute. D.C. Corbin's railroad received free telephone service in exchange for right-of-way and transportation rights for a velocipede for repair purposes.

In those early days, there were many problems launching a new business and telephone service was one. Many businessmen agreed with some points of view expressed editorially in the *Spokane Falls Review* during this period:

> The usefulness of the institution is questionable to us, for it seems rather queer to use a telephone in a place where a small boy can carry a message to any quarter of the business part of the city in two minutes On account of new poles being put up on Howard and Sprague Streets, stretching new wires both telephone and electric light, the wires have been mingled together during the past two or three days, and the consequences were that when attempting to call up somebody through the telephone you invariably hailed someone else.

Spokane's Rapid Development

Spokane was now developing rapidly. An article by Eugene V. Smalley in the *Northwest Magazine* of October 1887, described the new spirit from the perspective of a romantic reporter as follows:

> In all the broad northern belt of new country, which reaches from St. Paul to the Pacific Coast, I know of no scene of rapid development which equals that presented by Spokane Falls today. There is no such striking spectacle of the transformation of a frontier village into a large town, with extensive manufacturing, commercial and railroad activities. Nor are there [such] solid business blocks and handsome dwellings. It has a bustling population recruited on the arrival of every train by a throng of energetic, quick-witted newcomers.
>
> The click of the trowel, the rasp of the saw and the resonant blows of the hammer make music over all the broad, forest-girdled plain through which the blue Spokane rushes and leaps on its swift way to the Columbia and

Burke, Idaho, in March 1904, looking up Main Street.

Pole yard in Cataldo, Idaho, 1904

forms a sharp treble to the bass of the roar of the cataracts that furnish the incomparable water-power for the wheels of many mills and factories.

This is the music that the western man loves best—the rattle and hum of varied industry, laying the foundation and rearing the superstructure of a new civilization. Gazing in stolid wonder upon this wonderful transformation scene stands the sullen blanketed Indian who, but a few years ago, looked upon the flowery Spokane plain as his choice and exclusive domain and upon the river as created by the Great Spirit to bring fish to his nets.

1888 Diary Continues to Chronicle Work in Northern Idaho

January
13th I go to Wardner. 38 degrees below zero. Spend evening playing Casino with Miss . . . **15th** At Mission. Thermometer frozen at 40 below. I am a fool if I am here a year from now. **16th** At Mission. 32 below in Spokane. Am told I will work in Spokane in spring. I won't kick. [This was a discouraging period for him and he was tempted to leave telephone work.]

February
7th Feeling rather blue. Take a little walk up in the mountains this P.M. **13th** Go to Wolf Lodge and into Blue Creek Canyon to repair line. **15th** Mr. Norman [head of phone company] came to see me today **18th** I tell Mr. Norman that I want to get where I can learn something. **27th** Get check from Mr. Norman for $11,975.00! Mistake in account. They find Jerome's body today. **28th** Study "square roots" tonight. **29th** Learn to play chess.

March
13th Still at Mission. Greatly tempted to go into ranching. **19th** Go look at Thompson's ranch. Will buy it if he will give me time on part of it. **24th** At Wardner digging holes for poles. Deposit $150.00 in Bank of Murray. **27th** Supervising men in Wardner raising poles and stretching wire.

Norman's Telephone System office in Old Mission, Idaho, 1889

Telephone office at Wolf Lodge, Idaho, 1890.

Wolf Lodge telephone test station in 1888.

April
19th Get job offer to be a watchman for $2.50 per day. Mr. Norman offers me $60.00 per month to work construction. **28th** Establish camp two miles above Wallace. **30th** Digging holes all day. Hire two more men.

May
6th Move camp to one mile below Burke. **16th** Help install two phones at Osborne. **21st** Put up switch board at Wardner. **24th** Put in new office at Murray.

November
13th Go to Spokane to continue to work for the phone company.

A year later, in November 1889, the *Spokane Falls Review* reported optimistically: "The telephone company has 160 boxes, of which 70 have been placed this year. There are now 600 miles of telephone wire tributary to Spokane Falls, of which 150 miles was built this year. Every mining camp in the Coeur d'Alenes is now reached by telephone."

December
3rd Work on electric light line. **9th** At Wardner all day. Have breakfast and dinner with Miss Frisbie.

1889 a Year of Optimism

By 1889, 10,000 men were working in the mining camps of northern Idaho. The Coeur d'Alenes were then producing more than one-fifth of the entire lead consumption of the United States. In Spokane the first Monroe Street Bridge was opened in 1889. It was destroyed by fire and replaced by a steel truss bridge in the early 1890s. The current bridge was completed in 1911 at a cost of $500,000. It was the longest concrete arch in the United States.

1889 had been a year of optimism. Statehood for Washington had become a reality. From its first settlement in 1872 and incorporation in 1881, the population of Spokane Falls had grown to nearly 20,000.

Wallace, Idaho, 1889, shortly after first telephone was installed.

Wallace, Idaho, at the corner of Sixth and Hotel Streets in 1904. Telephone company office on left and opera house on the right.

The *Spokane Falls Illustrated* (published by Frank L. Thresher for the Spokane Falls Board of Trade) described the city as follows:

> Nestling among the pine-crowned hills on the banks of a river whose water power is the marvel of the known world, at the western extremity of a picturesque valley that opens up its broad bosom to the eastward until it gradually diminishes and is lost in the blending of the mountains and the heavens, rises the splendid young city of Spokane Falls . . . It stand peerless as the hub of the greatest silver-lead producing country known . . . Throughout this whole stretch of country new towns are budding into life . . . A great metropolis has risen in the midst of what was a primitive wilderness.

1889 Diary Entries

January
1st Charles and I get $10.00 toilet case for traveling purposes for Mr. Norman. **3rd** Talk with Miss Frisbie this evening. **7th** Lonesome and blue. About sick of everything. **27th** At Wardner all day. Have dinner with Miss Frisbie.

February
2nd Go to dance at Junction. Have a good time.

April
10th Take music lesson.

May
7th Begin work on Mullan line. **30th** One of the men falls off car and we run over him. He is hurt considerably.

June
7th Mr. Norman wants me to take charge of Wallace office. **14th** In Spokane all day. Go to Circus. See Miss J. **20th** At Wallace tending office. Mrs. Wallace goes to Murray.

August
4th At Wallace. Get word that Spokane is burned up.

Spokane Falls, looking east, after the 1889 fire which destroyed the entire downtown area. In addition to photographs in Thomas Elsom's albums that he had taken were ones widely available at the time, such as those of the 1889 Spokane fire. Elsom was in Wallace, Idaho, at the time of the fire.

This photo shows the guarded workmen cleaning out the vault of the First National Bank Building following the 1889 fire. Obviously, the telephone office and exchange, located inside the bank building, were destroyed. Note the bottom photo on page 39 for a view of this building before the fire.

Spokane's Great Fire

The *Spokane Falls Review* described the Great Fire as follows:

It originated at 6:15 P.M. in the roof of a lodging house on Railroad Avenue ... A dead calm prevailed at the time and spectators supposed the firemen would speedily bring the flames under control. This could have been done if proper precautions had been taken. But the superintendent of the water works was out of the city ... men in charge failed to call for more pressure.

The heat created a current of air and in less than half an hour the entire block of frame shacks was enveloped in flames ... igniting several adjacent blocks at the same time. Opposite the block ... stood the Pacific Hotel, one of the handsomest structures in the Northwest. It was soon ablaze and by that time a high wind prevailed from the southwest. It was evident that the entire business portion of the city was in danger. Mayor Fred Furth ordered that buildings be blown up with giant powder to check the spread of the fire ...

The Grand Hotel, the Frankfurt block, the Windsor Hotel, the Washington block, the Eagle block, the Tull block, the new Granite block, the Cushing building, the Falls City Opera House, the Hyde block, all the banks and in fact every house between Railroad Avenue north to the river and from Lincoln Street east to Washington Street, with the exception of a few buildings in the northeast corner, were totally destroyed.

Meanwhile, a sudden change in the direction of the wind carried the fire southward across Railroad Avenue and destroyed the Northern Pacific passenger and freight depots and several cars. The freight depot was a mammoth structure and was filled to the roof with valuable merchandise, a very little of which was saved.

Another eyewitness account stated:

The next morning the sun rose on a smoking, ruined city. The city was under military rule and no curiosity seekers were allowed to go into the burnt district. All saloons, unburned, were ordered closed. The citizens began to wander down the ruined streets, made ghastly by the great skeleton walls. The banks are all huddled into the Crescent Block and they each have a sign made hastily. The business firms are getting tents up to do business in.

It was estimated that 800 businesses and private residences in 300 buildings were destroyed and 2,000 residents were displaced.

The rebuilding of the city progressed rapidly. The tents of the first few weeks after the fire began to give away to the new, impressive edifices of granite and brick.

The fire destroyed the Spokane telephone system. Norman and Hopkins sold half of their interest in Inland Telephone and Telegraph Company to Sunset Telephone Company, which was the Bell system in the Pacific states. After the fire, the Inland Company set up its office in the new Hazel Block where they remained until 1892, when they moved to their new building on Wall south of Main.

The 1889 Diary Continues
September
3rd Sprain ankle jumping off train.

October
3rd Railroad people fighting for right of way give me some trouble.
10th Running twelve-man crew repairing telephone lines in 4th of July Canyon.

December
31st Finish 4th of July Canyon duties.

1890 Diary Entries
January
1st At Wallace.

July
1st Leave Wallace for Spokane.

August
1st Salary increase to $80.00/month

After the fire, the town of Spokane set up businesses in tents until they were banned in 1890.

The *Daily Chronicle's* temporary location after the Great Fire of 1889.

Telephone and telegraph offices shortly after the fire. The piled wire was probably used to replace burnt lines. This same building also served as the "Electric Light" facility. Note the lettering.

Adjusting fire loss claims following the 1889 fire.

Elsom's Daughter Recorded His Memories

Although the diaries for the years 1891 through 1896 are missing, insight into Thomas Elsom's business and personal life were available through other sources. The following quote was attributed to Elsom's daughter, Gertrude Ladd, who had listened to him and recorded his memories in his retirement. However, portions of this narrative are identical to the *Pacific Telephone Magazine* article in June 1931 (excerpts quoted earlier). It is difficult at this late date to determine which was the original source. In any case, the account is informative and colorful.

> During the intervening forty years [working for the telephone company], the story of Mr. Elsom's experiences and adventures reveals the history of an empire in the building, and tells of telephone progress in the Inland Basin between the Cascades and the Rockies. His hobby through all these years was photography, and among his collections of photographs are many pictures of early day telephone equipment and construction scenes that can be found is few other collections. If you listen to him, you learn of the time spent in the Coeur d'Alene mountain mining region of Idaho, converting the old army telegraph line purchased by Mr. Hopkins into a telephone circuit and installing telephone service in the rough mining-camp towns in the midst of the traditional dancehall-gambling house surroundings of the frontier; how he stole a ride in the bucket of an aerial cable tram that became stuck and held him suspended for hours, unknown and unmissed, above a deep gulch; how he prospected the route for the toll lead from the Coast over the Cascade Mountains to the interior, tripping when coming from the summit, and sliding down a 500-foot cliff, landing at the entrance of the Northern Pacific tunnel, which marked the route finally followed; how on one mountain pack train trip, with C. J. Corcoran, Mr. Corcoran decided to give his horse a rest and dismounted only to see the horse, at the next step, plunge down a 200-foot ravine to its death. For more than forty years, he blazed the routes and placed the leads that now traverse the mountains and rolling areas of Eastern Washington and Northern Idaho.
>
> In 1891 Mr. Elsom married Nellie Pratt, who was one of the first telephone operators. She gives an interesting account of some of her experiences, saying, "When I first entered the employ of the local Telephone Company, it was a Sunday morning in 1890. I had previously been employed in the Rochester, New York Company. The local operator was going off duty at 8 A.M. There were some 78 subscribers, all the calls being made by name,

Thomas Elsom, in 1889 at age 24.

Nell (Pratt) Elsom, in 1891 at age 22.

Thomas and Nell Elsom's first house, on Fourth Avenue in Spokane.

Thomas and Nell on their wedding day, October 17, 1891.

but the numbers were listed by the names, so it was easy to operate. The office was located in the rear of a one-story building, fronting on Stevens Street, where the Old National Bank now stands. The Western Union Telegraph Company occupied a part of the building. This was just after the fire in August 1889. Our office consisted of the business office in front and the toll lines department in the rear. There was a sky-light in the ceiling for light and ventilation, and one booth for public use. There were two operators employed during the day and one all night.

The old iron wires were hard to hear over. One amusing incident occurred when a call came from the Coeur d'Alene country for a Spokane Indian. A messenger was sent and a big brave came. We told him how to use the phone in the booth and the operator called, "Go ahead, Wallace, here is your Indian." It was necessary to holler to make Wallace hear, but the poor Indian in the booth got the full benefit and he burst out of the door of the booth, and went out muttering, "Debbil in there." We tried to get another Indian to come, but we could not. The Chinamen used the phone a good deal and they jabbered, seeming to have no difficulty.

Another amusing incident occurred. A call came for a prominent judge. It was very warm. The booth, or sweat box as we called it, was extremely torrid, but we ushered him in and shut the door, thinking he was accustomed to using the phone. After making connection, but listening in, we could not hear our learned judge talking, so the operator opened the door of the booth and saw the judge with the wrong end of the receiver to his ear, the sweat was pouring down his face. He was swearing terribly. At another time, a call came from a branch bank to the main bank here to send a certain amount of currency, but in transmitting the message, the operator mistook "currency" for "Thursday." When we found out our mistake, we were much concerned as the manager of the local bank was very crabby. However, we got ahold of one of his clerks and explained the mistake. He assured us he would straighten out the tangle, which he did, much to our relief.

Family Life

Thomas and Nell (Pratt) Elsom were married on October 17, 1891 and spent the next 62 years together. They had seven children: Russell Lincoln (born April 2, 1896), Gertrude Mary (born November 28, 1897), Joseph Lawton (born April 25, 1900, died in infancy), Thomas (born about 1902, also died in infancy), Floral Honor (born October 31, 1903), Ruth Jane (born January 21, 1904), Burnard Henry (born May 30, 1912).

A composite of the Elsom family with five of the children. Top row from left, Nell, Joseph, Thomas. Bottom row from left, Russell, Gertrude, Floral, Ruth.

Left: The Spokane National Building, where Elsoms had their first apartment after their marriage in 1891. **Right**: Interior of their home in 1903 at 1830 Summit showing Thomas's desk.

Left: Elsoms' home at 1830 Summit where they lived from 1893 to 1907. Right: Their home at 20 East Sixth where they lived from 1907 to 1911.

After their marriage, the Elsom's occupied an apartment in the Spokane National Building. In 1892 they moved to an apartment in the Dodd Block. The following year, they moved to a house on Fourth Avenue east of Stevens. Later the same year, they moved into a home at 1830 Summit Boulevard, overlooking the Spokane River and Peaceful Valley. It was located several hundred yards west of the Maple Street Bridge, making it very convenient for Thomas to walk to his office at Wall and Main. The home was later extensively remodeled by Nellie's brothers, who were building contractors. They built medium-priced homes during that early period of growth in Spokane and then continued the business in Seattle. Their third child, Lawton, died there. Unfortunately, this beautiful place had to be sold to the railroad in 1907 to make way for the new right-of-way. The old basalt rock foundation still remains, but the house was apparently relocated to a site on Sharp Avenue, south of the Holmes School in the vicinity of the old Natatorium Park entrance.

They next moved to a home at 20 East Sixth Avenue, where they lived for three years. One of their neighbors was Peter Jacoy, who owned the well-known cigar store at Sprague and Washington. They had a lot of sickness there and, sadly, their fourth child, Thomas, died there in infancy of scarlet fever.

In 1911 Elsom quit the telephone company for two years and bought an 80-acre farm with a small home at Saltese, which they called "Valleyview." The modest house was set in a ravine among pine trees that whispered in the wind. They remained at this location until a couple years after Elsom's retirement from the telephone company in 1930.

During his long career with the telephone company, Elsom continued to work his way up through the ranks of the growing telephone business. He supervised various facets of the construction, development and expansion of the telephone system in the Inland Northwest, contributing his own inventions, designs and modifications to further the whole process. Much of his work is detailed in the next chapter. The following photographs, taken by Elsom, depict telephone expansion in Spokane in the 1890s.

M.Monler, early Spokane telephone installer who worked for or with Thomas Elsom.

The Spokane telephone line crew in 1899.

Telephone crew raising a pole in Spokane in 1899 without the aid of hoisting equipment, which Elsom designed later that year.

Spokane Exchange switchboard in 1898.

Battery room for the Spokane telephone exchange.

Elsom and workers on the phone lines, looking north on Howard at Riverside. The Holley-Mason Building was built in 1890.

Inside the telephone office of the Spokane exchange in 1891.

Miss Hodge at the switchboard of the Spokane telephone exchange in 1899.

Form 408. 3-15-99-50,000.

THE AMERICAN BELL TELEPHONE COMPANY.

SPECIAL DIRECTIONS
RELATIVE TO SOLID BACK TRANSMITTERS.

NEVER HIT NOR **VIOLENTLY SHAKE** THE **TRANSMITTER.**

NEVER ATTEMPT TO **ADJUST** THE **TRANSMITTER** NOR **LOOSEN** THE **NUTS** OR **SCREWS HOLDING THE BUTTON** IN PLACE.

THE MOUTHPIECE AND NICKELED PARTS OF THE TRANSMITTER SHOULD BE CLEANED AT EACH INSPECTION; IN DOING THIS, **BE CAREFUL NOT TO PRESS UPON THE DIAPHRAGM** NOR UPON THE NUTS ON THE FRONT OF THE DIAPHRAGM.

IT IS DESIRABLE, FOR EXCHANGE SERVICE, TO USE A STRONGER BATTERY WITH THE SOLID BACK TRANSMITTER THAN IS USED WITH THE BLAKE TRANSMITTER. WHEN USED IN CONNECTION WITH **LONG DISTANCE LINES,** IT IS RECOMMENDED THAT BATTERY BE USED WHICH WILL, AT ALL TIMES, GIVE A **CURRENT OF MORE THAN .42 OF AN AMPERE** AT THE END OF ONE MINUTE AFTER THE BATTERY HAS BEEN DISCONNECTED FROM THE TRANSMITTER AND CONNECTED IN CIRCUIT WITH AN AMMETER AND A RESISTANCE OF FIVE (5) OHMS.

IT IS SUGGESTED THAT SUBSCRIBERS BE CAUTIONED:

NEVER, UNDER ANY CIRCUMSTANCES, TO **HIT** OR **TOUCH** THE **DIAPHRAGM** OR THE **NUTS** UPON THE FRONT OF THE DIAPHRAGM OF THE TRANSMITTER.

WHEN TALKING TO **PLACE THE LIPS CLOSE TO THE MOUTHPIECE** AND **SPEAK IN A FIRM TONE** OF VOICE.

Special directions for use of telephone in 1899.

A SUPERINTENDENT OF CONSTRUCTION—1898

Thomas H. Elsom in his office at Spokane in 1898, when barbers did not do a rushing business.

A news clipping of Thomas Elsom at work in his office in Spokane.

An early 1900s vintage "White" truck equipped to install telephone poles.
This method was a major advantage over placing poles by hand.

GENERAL INFORMATION.

Subscribers' Numbers.—Each subscriber's telephone is designated by a number placed to the left of his name. The number is prefixed by the name of the exchange with which the subscriber is connected. In calling for subscribers, it is necessary that both the exchange *name* and *number* be given.

Instructions for using Telephone.—In talking, speak in a moderate tone and directly into the transmitter, with the lips as close as possible to the mouthpiece.

To call Central Office, take the hand telephone from the hook, place to your ear, and operator will say, "Number?"

When operator receives exchange name and number of subscriber wanted, she will repeat them to you in order to avoid mistakes.

Remain with telephone to your ear until party called for has answered.

When a conversation is finished, be sure to replace the hand telephone on hook in original position.

How to answer a Telephone call.—Remove the hand telephone from the hook and say: "Here is Main 297" (or whatever your number may be). The party calling should say: "This is Main 298 (or whatever the number may be). Much friction and annoyance will be avoided if this simple plan is carried out.

Non-Subscribers.—When it is desired to have a non-subscriber called into one of our public offices at a distant place, a messenger will be sent for at the expense of the party ordering service.

Non-subscribers can have the same service as regular subscribers by application at any of our public offices and paying our established rates.

This is a page from the January 1898 Telephone Directory for the states of Washington, Oregon and Idaho.

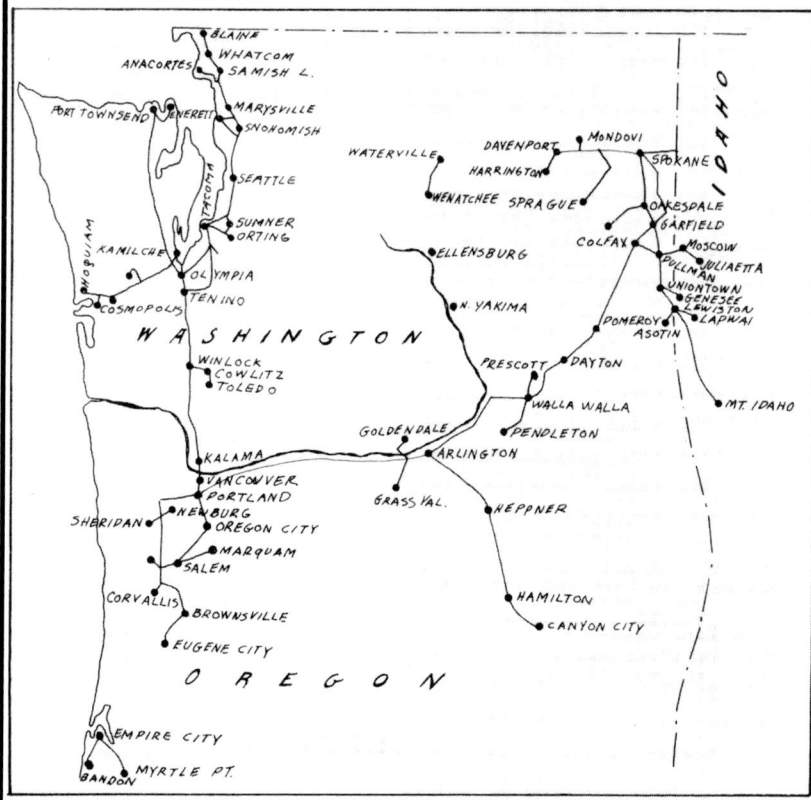

This map was taken from the January 1898 Telelphone Directory for the states of Washington, Oregon and Idaho. Total subscribers at that time for the region covered was 7,368.

Chapter 3
Inland Northwest Telephone Expansion in Thomas Elsom's Words

The following early history of the beginning and growth of the telephone system in the Inland Empire is from an account written by Thomas Elsom in 1931 covering the expansion through the year 1915. It was later typed by his oldest daughter, Gertrude Elsom Ladd, the mother of this compiler. With minor editorial changes, the history is presented as Gertrude Ladd typed it. It is followed by more excerpts from Thomas Elsom's diaries to offer greater insight into his personal perspective and experiences:

Owing to the experimental stage of the telephone business in general, the lack of knowledge regarding proper construction methods, the undeveloped condition of the country, indefinite location of roads, and the general failure to anticipate the rapid development of the telephone business that was just beginning, the construction and equipment of the system up to this time was necessarily of a cheap and temporary nature.

In 1891 and 1892, the business had developed to such an extent in the Inland Empire and in the Oregon Territory, of which Portland was the center, that it became necessary to establish a trunk line between these two commercial and telephone centers. Also, an ambition began to develop for telephone communication with San Francisco. In order to make this a practical success, it was necessary to reconstruct practically the entire system along more modern lines, which had been pointed out by experience and study up to that time.

The Spokane-Portland Lead

Accordingly, a study was made and a route selected that promised to be the most permanent and direct one between Spokane and Portland. This route followed the same general course as the existing line between Spokane and Walla Walla, namely via Spangle, Garfield,

Colfax, Almota, Pomeroy and Dayton. Owing to the establishment of roads and the inadequacy of the existing lead, a reroute of a large portion of the line was made and now 25-foot red cedar poles with six-pin arms were placed with a loop of 172# copper for the through business. The old wire was retained for local business.

From Walla Walla via Wallula and Umatilla to John Day, which was then the dividing line between the Inland and Oregon territory, a new lead of 25-foot and 30-foot poles with six-pin cross arms and 172# copper loop was constructed in accordance with the best methods known at that time.

Along the Columbia River, there was no wagon road, so it was necessary to build a barge and hire a steamboat for the transportation of materials and camp accommodations. For a considerable distance along the river, it was necessary to carry poles and material by hand from the river up the bluffs, which were so steep that teams could not be utilized. Poles were rafted along the river from railroad stations to the nearest point of use. This construction was met by a similar construction of the Oregon company. The telephone connection was established in 1892 between Spokane and Portland, and shortly after extended on to San Francisco.

In 1893 a general depression of business began that lasted for three or four years, retarding the development of the telephone business. About 1896 and 1897, the conditions began to improve and it was necessary to resume the extension of the telephone system to meet the demands for service at various points and in 1898 the toll line was extended from Walla Walla to Baker City and Huntington thus making connection with the Rocky Mountain Bell Company. This naturally boosted the business over the Spokane-Portland lead in both directions from the junction point at Walla Walla, necessitating additional facilities along the Spokane-Portland lead which were provided by the addition of a 435# cooper loop between Spokane and Portland and several short loops along the route to care for local business between busy points and thus relieve the original through circuit for handling through business

Dayton-Starbuck Lead

In 1900 a branch of the Spokane-Portland lead was constructed from Dayton to Starbuck consisting or 25-foot red cedar poles and l2# iron loop. This feeder was built to furnish service to the town of Starbuck, which had developed due to the establishment by the railroad company, of shops at that point, and also to supply service to a large stock ranch along the route, the owner of which was very anxious for the service and afforded good prospects for a paying business. The company also had in view the extension of this line to Riperia, which was the transfer point between railroad and boat transportation. This extension was made later. At the date of this inventory (1915) this lead still consists of 25-foot poles carrying one #12 iron circuit on brackets.

Mayview Branch

In 1901 a small rural town called Mayview had developed a few miles south of Almota and the people of the community making a strong plea for services and prospects of a paying business being good, a branch of 25-foot red cedar poles and #12 iron loop was constructed from the Spokane-Portland lead to supply them with service. At the date of this inventory (1915), this lead still consists of 25-foot red cedar poles carrying one #l2 iron circuit on brackets.

Reinforcement of Spokane-Portland Lead in 1901

The construction of these feeders, together with the development of the business in other directions and a general appreciation on the part of the general public of the advantages afforded by the telephone to facilitate the handling of business, caused a continual demand for the service in every direction, and consequently about 1901, it was necessary to do extensive work on the main lead from Spokane to Portland. This work consisted of the placing of additional wires and cross arms and a general strengthening of the pole lines by replacing weak poles and guys and placing additional poles. This work was especially heavy along that part of the lead between Spokane and

Walla Walla, where the number of wires had gradually increased to such a number that the lead far exceeded that for which the pole line was designed. From the year 1902 to 1908, the increase in the business was very gradual, during which time several projects of minor importance were carried to completion that were necessary to care for the additional demand for service.

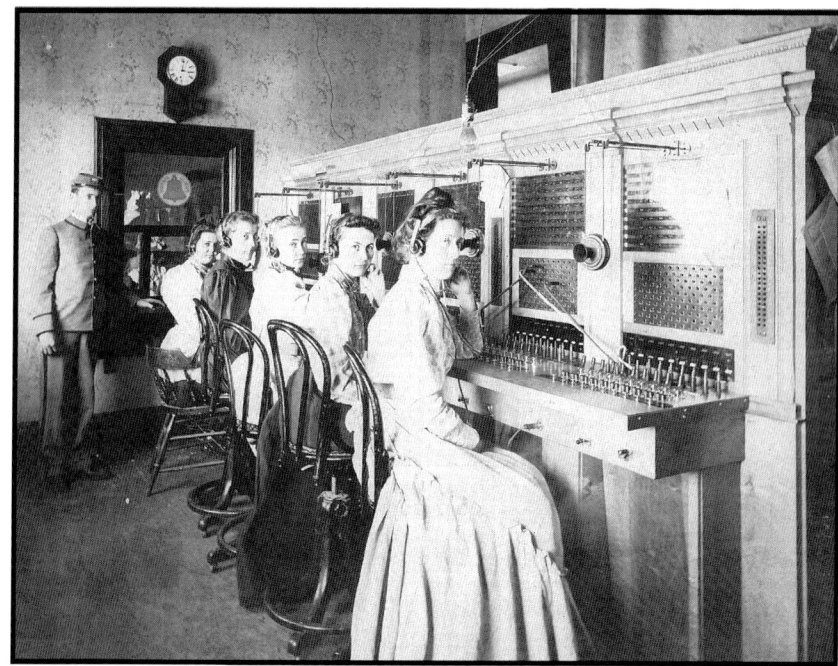
Telephone exchange in Walla Walla in 1898.

During 1908 that portion between Walla Walla and Arlington was entirely reconstructed by replacing defective poles, rerouting line in order to conform with new location of roads, etc. In 1909 the lead between Walla Walla and Spokane needed extensive repair work in order to place this lead in good physical condition to withstand the additional circuits, which were recommended by the traffic department. By 1910 the entire lead was reconstructed consisting of rerouting, placing approximately 1150 poles carrying two 10-pin arms and 4 circuits of 172# copper and one circuit of 435# copper.

In 1911 additional circuits were necessary between Walla Walla and

Athens. These additional circuits made it necessary to do extensive reconstruction work between Walla Walla and the Oregon state line, and on to Pendleton as this line had not been worked on, except for light repair, since its original erection in about 1898. A branch line from the main Spokane-Portland line from Walla Walla to Prescott and Eureka was purchased from a Mr. Isaacs for $2,000. This line had been constructed by Mr. Isaacs to provide service for Prescott and Eureka, which are located in a rich wheat belt. Owing to the non-importance of this line no extensive reconstruction work has been performed since the original purchase.

Spokane-Seattle via Wenatchee

About 1887 the territory west of Spokane, commonly called the Big Bend country, began to come into prominence as an agricultural and stock section, and with the development of other lines of business then arose a need of communication. This was especially desirable at this time due to lack of railroad facilities, the supplies being obtained from Spokane by freight teams, and mail and passengers by stage. The principal source of traffic being from Spokane, west through Deep Creek and Davenport to Fort Spokane on the Columbia River. The government had previously established a telegraph line from Spokane to Fort Spokane. It was little used for commercial purpose, resulting in the acquirement of the line by Mr. Hopkins and installation of telephone instruments at Deep Creek. At that time, Deep Creek was a lively little rural town due mainly to the fact that there was a small flour mill at that point where the farmers brought their wheat and had it converted into flour. There was a small supply point between Deep Creek and Davenport called Mondovi, where a telephone station was established. Davenport was a growing town of some importance as a stage and freight station and natural distributing center for considerable country. A telephone station was established there and also one at a little place called Egypt between Davenport and the fort.

This line consisted of pine and tamarack poles from Spokane to a point near Egypt for a short distance from the fort and a stretch of 10 or 12 miles in the vicinity of Egypt. On account of the practical abandonment of the fort, the line between there and Davenport was a short time later dismantled.

This country slowly settled up and developed and extensions to the telephone system were made as the business demanded. About 1894 it was necessary to reconstruct the lead between Spokane and Davenport in order to properly care for the growth of the business. By this time the railroad had been extended from Spokane into this country and the towns of Wilbur, Creston, Hartline, Coulee City, etc., came into existence.

Town of Wilbur on August 10, 1898

About 1898 the line was extended from Davenport to Wilbur consisting of 25-foot red cedar poles and 172# copper loop. In 1899 extension was made from Wilbur to Coulee City. It was necessary to place an additional loop between Spokane and Davenport. In 1900 the line was extended from Coulee City to Waterville with 25-foot red cedar poles, cross arm and 172# loop. Owing to lack of railroad or other shipping facilities, it was necessary to haul the poles and other material and supplies used on this extension for the entire distance of about 45 miles by team. This method of hauling was very expensive on account of the crossing of two deep coulees, with long steep hills on each side.

In 1893 a line was built from Waterville to Wenatchee to furnish communications between Waterville and its shipping points at Orondo and Wenatchee on the Columbia River.

Thus upon completion of the line between Coulee City and Waterville in 1900, connection was completed from Spokane to Wenatchee, which due to the fruit industry of the Wenatchee valley, was developing into a business center of considerable importance, this together with the fact that about this time the system had been extended north from Waterville to Pateros, Twisp and Brewster, made necessary extensive work along the lead from Spokane to Waterville in order to provide required facilities for properly handling the increased business.

About this time, Seattle began to be an important marketing place for the produce grown is the vicinity of Wenatchee, and also numerous people of Seattle became interested in mining and other projects throughout the Chelan and Okanogan districts north of Wenatchee. These conditions produced a strong demand for telephone connection between Wenatchee and Seattle.

The service between Spokane and Seattle was also in need of improvement on account of the roundabout way, and the usual congested condition of the lines over which it was necessary to carry the business. In consideration of these conditions, in 1901 the line was extended from Wenatchee to Cle Elum, making connection there with a line that had been extended to North Yakima from Seattle in 1900. In this way, direct connection was made between Spokane, Wenatchee and Seattle.

In 1902 it was necessary to place another circuit between Spokane and Seattle, and further relieve the existing wires by the addition of short trunks between busy points to care for the business of a local nature. In the construction of the line between Wenatchee and

The L.V. Wells Drug Store at Wenatchee, Washington, with a new telephone cable pole in front, 1888.

Telephone office at Peshastin, Washington, in 1910.

Cle Elum, unusual conditions were encountered due to lack of shipping facilities for a distance of about 40 mile over the mountains. It was necessary to haul poles and material by team over country where there was but little travel and in places it was necessary to repair or construct sections of road to get through with our teams.

Between 1902 and 1905, the demand for service from Spokane to Seattle was increasing so rapidly that in order to provide facilities and relieve the congested circuits, an additional circuit of 435# copper was strung from Spokane to Cle Elum and on to the west end of the Stampede Tunnel where it connected with the circuit strung from Seattle.

The placing of this through circuit necessitated replacing the present 6-pin cross arm between Spokane and Deep Creek with a 10-pin arm. Also through various sections between Hartline and Waterville, it was necessary to place additional poles in order to half space the line on account of the occurrence of very heavy frosts and sleet, which would break the wires on long spans. The poles in this section were half-spaced for a total distance of approximately 30 miles. In 1907 congested circuit conditions required additional service from Spokane to Coulee City to care for local business.

In 1909 it was decided to reroute a portion of the line in the vicinity of Wenatchee because of private property encroachment and deterioration of the line. This made approximately five miles of new line, which included the crossing of the Columbia River; however, these five miles replaced the eight miles of deteriorated line that occupied private property. At this time, a joint interest was obtained in the pole line between Wenatchee Junction and Wenatchee (125 poles), from the Farmer Telephone and Telegraph Company for a sum of $7.00 per pole. During the year 1910, demand for additional toll facilities on the Seattle-Spokane route was exceptionally heavy and more improvements were necessary. In 1912 and 1913, much work was done to take care of the ever-increasing demand for service.

About 1889 a branch line was built from Deep Creek on the Spokane-

Seattle lead south to Medical Lake when the demand for the telephone service had been created by the establishment of a state hospital for the insane at that place. Medical Lake was becoming quite a resort since it was near Spokane and the water [was thought to have] medicinal qualities. The line consisted of 25-foot red cedar poles and single-iron wire. This line was shortly after extended to Cheney, a promising town on the main line of the Northern Pacific Railway. About 1890 the town of Sprague about 25 miles west of Cheney and at that time the county seat and also division headquarters for the Northern Pacific Railway, began to ask for telephone service, but as the route was through rough and rocky country with little prospects of intermediate business, the construction was delayed until 1892, when the extension was made consisting of 25-foot red cedar poles with a single 172# copper wire.

The country along the Northern Pacific Railway gradually developed and in 1899 the telephone line was extended from Sprague to Ritzville with 25-foot red cedar poles and #9 iron wire. An extension was also made from Wallula to Pasco. In 1900 extension was made from Ritzville to Lind. In 1901 an additional wire was placed along this lead and the circuit changed from ground to metallic.

About this time the irrigation system in the Yakima Valley was being extended towards Prosser and Pasco and considerable demand was created for telephone service through this new section. There also was a demand for connection between the Yakima country and Walla Walla, consequently in 1901, the line was extended from Yakima to Prosser and in 1902 from Prosser to Pasco, and additional wire placed between Pasco and Walla Walla. The business continued to increase and also the demand for better service between Spokane and the Yakima country, and in 1905, the line was extended from Lind to Pasco making connection there with the line that had already been constructed between Pasco and Yakima. At the same time, the line between Spokane and Cheney was rerouted and old iron wire replaced with new copper, and additional circuits placed between Spokane and Ritzville, thus providing a more direct route and much improved service between Spokane and Yakima country, and at the same time, providing an auxiliary route that might be utilized for business between Spokane and Seattle in case of failure on the regular Spokane-Seattle route via Wenatchee.

The line between Spokane and Yakima crossed the Columbia River at Pasco on 85-foot poles. These poles failed to give proper clearance over the river after larger boats began to navigate. In order to eliminate this condition, it was deemed advisable to place special angle-iron fixtures on the Northern Pacific Railroad bridge. In 1907 a contract was given to the Union Iron Works of Spokane for placing 18 special angle-iron bridge fixtures for the price of $1628.00.

Waterville-Pateros-Molson Lead

In 1900 the country north of Waterville including the Okanogan Valley and the Lake Chelan district had developed considerable business due to the growth of the fruit and agricultural industries, promising mining prospects and also the development of the summer resort business on Lake Chelan. There were no railroad facilities above Wenatchee and transportation was by boat and teams, consequently mail service was not of the best and the need of telephone connection was strongly felt, so in this year, the company extended its line from Waterville to Chelan and Brewster. The poles were shipped by rail to Wenatchee and then transferred to boats and taken up the Columbia River to Chelan Falls and other landing places from which they were hauled and distributed by team. In this same year, British Columbia parties namely "Donald and Davis" constructed a line from British Columbia south to Brewster. This lead was cheaply constructed of native poles. This line was routed via Loomis, Conconully and Okanogan with a branch to Riverside.

Pateros-Twisp Branch

In 1901 a branch line was built from Pateros to Twisp of 25-foot Red Cedar poles and single #9 iron wire. The demand for this circuit was due to development of fruit and farming industries and work on mining prospects. Little change was made on this line up to 1915.

Dyer-Bridgeport Branch

In 1901 extensions were made to Dyer consisting of 25-foot red cedar poles and single #9 iron wire. In 1902 this branch was extended to Bridgeport with the same construction. The demand for this line was created by the development of the farming industry in that vicinity.

New telephone office in Lewiston, Idaho, April 1906.

Rosalia-Lewiston Lead

About 1886 a cheaply-constructed branch of the C. B. Hopkins telephone system was extended from Colfax to Pullman, Moscow, Uniontown, Lewiston, Genesee and Palouse. Poles from 15 feet to 25 feet were used with iron wire. That section which is now a very productive farming district was then new and in the course of development. About this time the Northern Pacific Railroad was building a branch road into the district from Spokane via Pullman to Uniontown and the Washington State Agricultural College established at Pullman, and the State University of Idaho at Moscow. The original construction soon became inadequate to the business, and about 1892, these lines were rebuilt and rerouted along more permanent routes.

In 1894 a line was built from Moscow to Kendrick and Juliaetta to handle business developed by extension of the railroad into that section. In 1897 an additional #9-iron loop was placed between Pullman at Lewiston. In 1899 still more work was done to improve the service. Garfield and Oakesdale were included in the service about this time, and improvements made until 1905, but no work of any magnitude was performed until about 1910. By this time, traffic was so congested it was necessary to do general reconstruction work.

Lewiston-Asotin Branch

This extension was made in 1897 to supply service to the prosperous little farming town of Asotin and consisted of 25-foot red cedar poles carrying a 4-pin cross arm with #9-iron wire.

Davenport-Odessa Branch

In 1897 the Great Northern Railroad had recently been built through the Big Bend country west of Spokane and stations established at Harrington, Lamona and Odessa. This large area of good wheat land developed into busy little rural towns from which large quantities of wheat were shipped and a strong demand developed for telephone connection with Spokane in order to furnish the wheat buyers with prompt and convenient communication with the managing offices of the different grain dealers who were located at Spokane and Seattle in most cases. To satisfy this demand, in 1897 the company built a line from Davenport to Harrington using 25-foot poles and #9-iron wire. In 1901 this line was extended to Lamona and Odessa with 172# copper loop. In 1902 the branch lead was built to Downs.

Edwall Branch

In 1899 a branch lead was built from the main Spokane-Waterville lead to Edwall with intermediate stations at Espanola and Wauken. This line was necessary to furnish service between the farmer and grain buyer. This is a productive grain-growing area. In 1900 the line was converted to metallic circuit because of need for improved service.

Hartline-Ephrata Branch

About 1906 a branch lead was constructed from the main Spokane-Waterville lead at a point near Hartline to Wilson Creek, a wheat and stock-shipping point on the Great Northern Railroad and in 1910 the line was extended to Ephrata. The demand for this extension was due to the development of the grain and stock business in this section and also business was created by the building of a sanitarium at Soap Lake.

Ferginson camp near Milton, 1910.

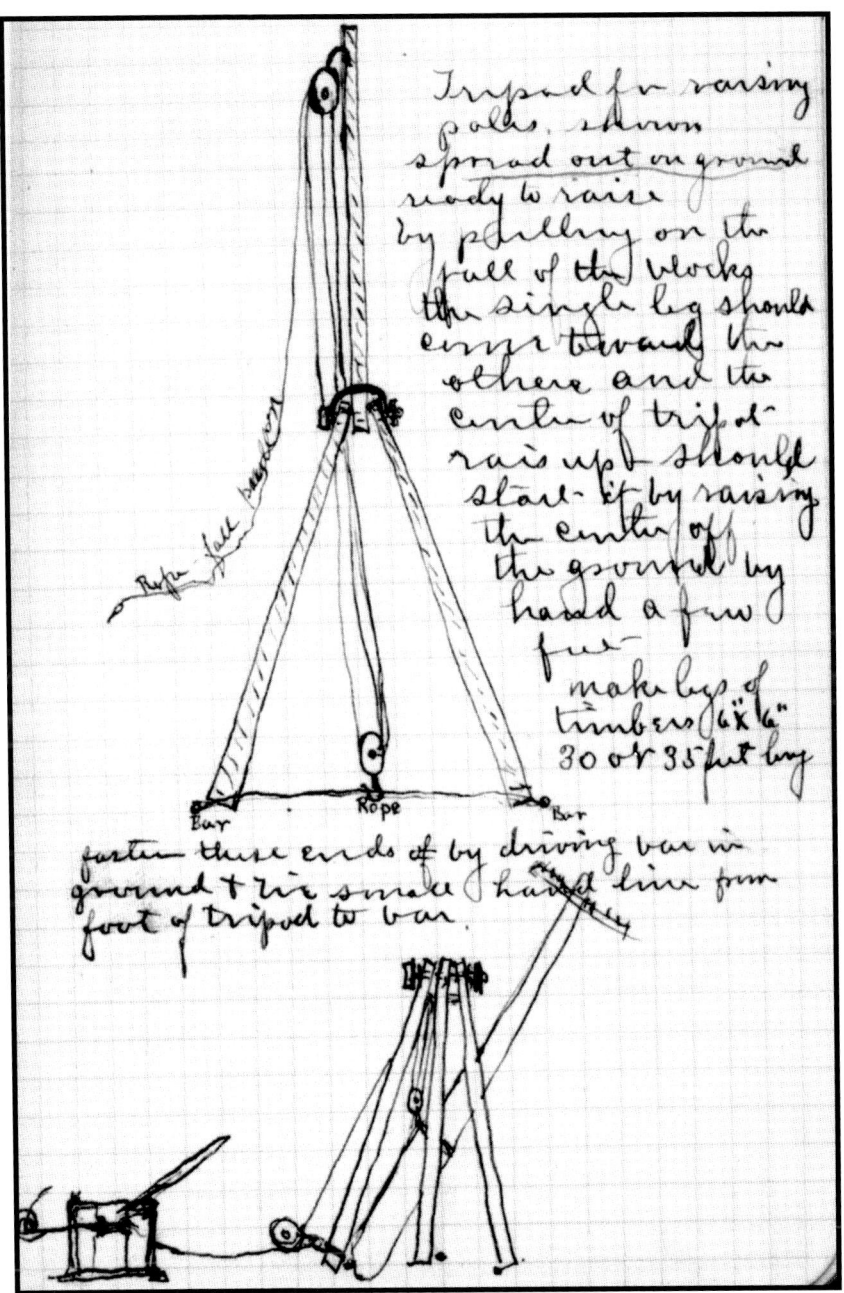

Tripod arrangement used to raise telephone poles and set them in holes. About six men tug and guide the pulley rope. (**This and the following schematic drawings are from Thomas Elsom's notebook.**)

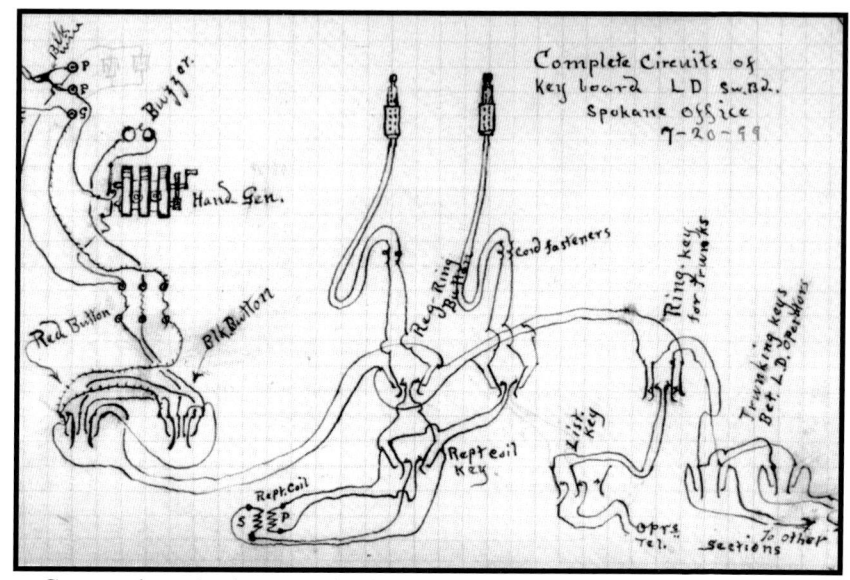

Composite telephone and telegraph circuits in the Spokane office, July 20, 1899.

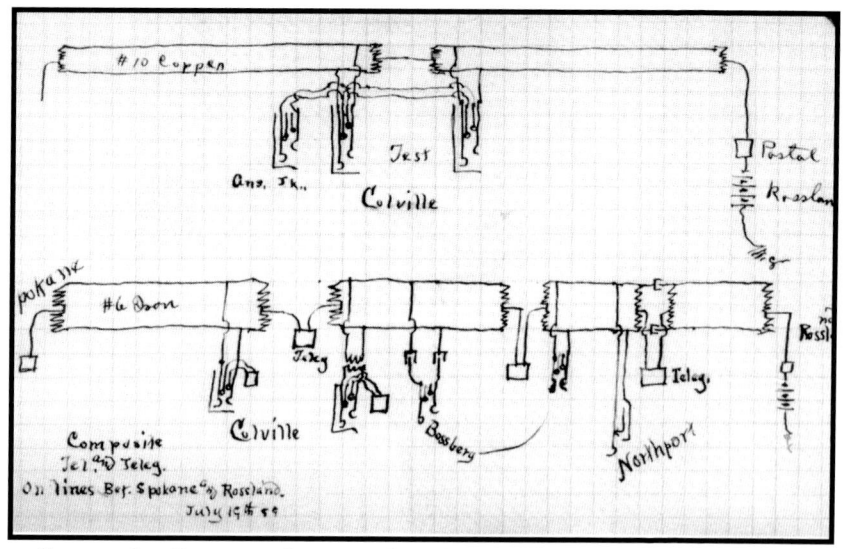

Composite diagram of connections between the Spokane office and Colville, Bossburg, Northport and Rossland, July 19, 1899.

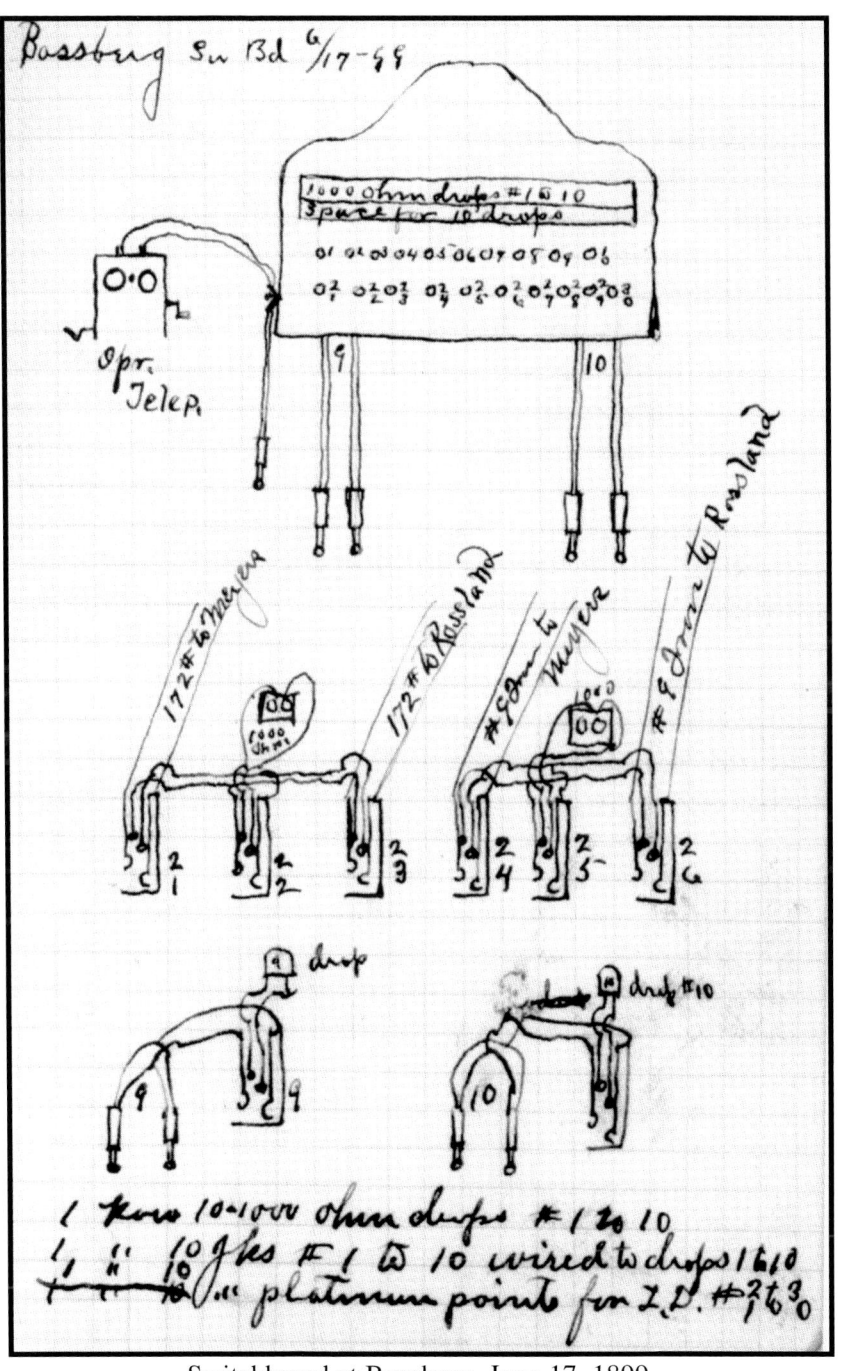

Switchboard at Bossburg, June 17, 1899.

1897 Diary Entries
November
1st Supervising phone company construction crew of eighteen men in Spokane. **4th** Supervising construction at Arlington, Oregon. **16th** Working at Pendleton, Oregon. **28th** At Pendleton. Get message from home that Gertrude [first daughter] born at 1:00 A.M. All OK.

December
25th Filing maps and drawings in Spokane office on Christmas day.

1898 Diary Entries
January
20th Manager from San Francisco in Spokane looking things over. **31st** In Spokane managing a fourteen-man crew.

February
24th Assist. Supt., supervising telephone construction in Idaho, Oregon and Washington. Walk from La Grande, Oregon to Union, Oregon. [About twenty miles.]

March
29th Take picture of crew putting up poles in La Grange. Rest of month supervising construction in Oregon and Washington. Became sick in bed.

April
1st Meet with Mr. Hopkins.

Elsom continues the same activity through June then gains the title of superintendent of construction for the Inland Telephone and Telegraph Company. He supervised the installation of telephone systems in many towns in Washington, Oregon and Idaho through 1899.

Diary Entries 1899
February
1st At Spokane office. When Armstrong gets in from Pomeroy send

him to Garfield and Palouse on the double. Sandburg finished putting in switchboard at Latah and Rockford. **11th** Receive new booths from Portland. Forward one to Republic. Set two in Hotel Spokane [at the location of the present Ridpath Hotel] and one in the Hazel Block. Took thirty pictures tonight. **20th** [This entry was regarding improvements made at Palouse.] The office is on the north side of the street and the pole line is on the south side. The lead divides in the middle of the block . . . **26th** Go to Meyers this morning by train. Go by team from Meyers to Kettle Falls this afternoon. Snowing a little and cold. **27th** Breakfast – Kettle Falls at 4:30 A.M. Go to Republic with team – have hard trip. Arrive about 6:00 P.M. Take Rohde over to start staking line. **28th** Leave Republic at 3:00 A.M. on stage for Meyers – have bad trip, don't feel well. Driver hurt one of his horses. Has lots of trouble with harnesses. Arrive at Kettle Falls at 3:30 P.M. Team had gone to train. Said we could not possibly make it and would take us no further. Hired another team. Six of us gave him $1.00 each. He ran the horses to the train a little late but we made it. Arrived in Spokane about 8:00 P.M. Left Rohde at Republic to start staking line. People say road is about 35 miles long from Kettle Falls – is not cleared wide enough for wagons – does not follow State survey all the way through. People expect to do more work

Republic, Washington, in 1898, was an early telephone site.

on road soon. Timber is very thick. Think would have been better to string wire on trees near ground. Is 2 to 4 feet of snow most of the way through. Crossing of Columbia at Kettle Falls is about 2500 feet.

March 1899
4th At Spokane all day. String Tekoa line ground out a mile further east. Seems to improve the working of line a little. Get pay check cashed and pay bills. Rohde finished staking line from Meyers to Republic – 1170 stakes from Republic to river – 138 stakes from river to Meyers. Bryant starts on Waverly line today.

[The following notation in Elsom's diary indicated the distance from Spokane to various railroad stations he used.]

> Spokane to Mead 9, Dragoon 19, Buckeye 24, Deer Park 27, Clayton 32, Olson Siding, Loon Lake 39, Lime Spur 43, Springdale 47, Valley 57, Chewelah 65, Addy 74, Arden, 81, Colville 88, Meyers 96, Marcus 103, Bossburg 111, Marble 117, Northport 130, Frontier 137, Sheep Creek 138, Silica 143, Rossland 147.

8th At Lewiston all day looking over town with Corcoran. Planning to rebuild exchange and making up estimate for materials. **12th** Sunday. At Spokane all day. Get Kringer started out with camp outfit on postal work. [This was the start of a large project. Kringer replaced Rohde as foreman who had quit over pay three days earlier.] Commenced my Dr. Munch treatment – Hot and cold foot bath and "fire glow."

[The following is a bill of provisions needed to set up camp for this new work.]

100 pounds flour -$1.85, one case St. Charles cream - 4.75, one sack potatoes - 1.85, one pound tea - .60, 50 pounds sugar - 3.45, ten pounds salt - .20. 25 pounds cabbage - .75, five pounds baking powder - 2.25, yeast cakes - .05, two cans pepper - .20, six bottles

mustard - .50, two bottles catsup - .60, one case corn - 2.25, one case tomatoes - 2.75, one box apples - 1.00, cinnamon - .10, one box nutmeg - .25, one bottle lemon extract - .25, one bottle vanilla extract - .35, one half case eggs - 3.25, 20 pounds butter - 5.40, 25 pounds bacon - 3.10, ten pounds dry salt pork - .90, one J&M coffee - .40, one package macaroni - .15, soda - .10, soap - .30, 25 pounds beans - 1.00, ten pounds lard - 1.00, five pounds rice - .40, 12 dozen matches - .25, ten pounds rolled oats - .35, two gallons maple syrup - 1.64, 15 pounds onions - .25, ham - 1.45, one gallon vinegar - .50, 35 pounds fresh beef - 2.88, one earthen jar, one coffee mill, two dish pans one funnel and a 12 gallon boiler tin.

15th Need 14 50-ft. poles on Division . . . to extend lead and about . . . on Indiana west. Go out to Kringer's camp this morning and return about 11:00 AM. They are out about 7 miles north. He has more men than he needs. Has 9 linemen. He also has 4 men tying with 1 man to carry up wire ahead for blocks. There are 2 sets of blocks with 2 men to each set. He has 1 man with solder outfit, 4 reel men, 2 men distributing material and moving camp. He also has 1 foreman, 1 assist. foreman, 3 teams and 1 cook. **19th** Sunday. At Spokane office this forenoon. At home this P.M. Klink arrives at Oakesdale this morning to take charge of gang to string loop from Garfield to Oakesdale and from Colfax to Pomeroy. He should have 4 or 5 linemen, 4 men on blocks, 4 men pull slack, 1 man solder and 2 men on teams. **20th** At Spokane office. Change Klink from Walla Walla . . . foreman at $2.50 /day and board. Change Hatley from Patrol #1 to Pendleton as repairman @ 60.00/ mo. Change Sherman from Spokane as groundman to lineman on patrol wagon with Ralph. **21st** Go to Garfield on train this morning. Stay with Klink's gang until 3: 00 P.M. and take train to Spokane. Arrive at 7:45 P.M. Klink has 1 foreman, 4 linemen, 4 reelmen, 2 on blocks, 1 to solder, 1 swamping and 2 teamsters and teams. It is cold, windy and snow. **24th** Go to Colfax by train this morning. Get Klink started out with new loop to Pullman and 204 6-pin cross arms at Colfax near elevator. 3 50-ft. poles there and 2 50-ft. poles down town. Ralph unload 65 exchange cross arms at Tekoa and ship 25 to Oakesdale. **27th** At Spokane all day. Get orders to build line from Lewiston to Grangeville.

Downtown Colfax during the flood of 1910.

The Colfax flood of 1910.

Downtown Colfax during the flood of 1910.

Downtown Colfax during the flood of 1910.

Pullman flood of 1910. Telephone office to the right.

Move office at Wallula and La Grande. **28th** Go to Lewiston by train. Stop overnight. Look around town some. Raining and cold. **29th** Leave Lewiston on stage for Mt. Idaho this morning. At 6:00 P.M. eat dinner at Waha – supper at Westlake. Get to Cottonwood about midnight and go to bed. Had a hard trip. Roads in bad shape with mud and snow – about 4 feet on hill use sleigh crossing mountains. Have to walk a good deal. One of the horses got down in the mud tonight. **30th** Leave Cottonwood 6:00 A.M. on stage. Get to ___ about 8:00 A.M. 8 miles – walk half way. Arrive Grangeville 11:46 A.M. Arrive Mt. Idaho 1:30 P.M. Get dinner. Walk back to Grangeville about 4:00 P.M. Talk about poles at Mt. Idaho and Grangeville. Can get Tamarack poles at Mt. Idaho as follows: Cut in timber at 1 cent/ft. Delivered in Mt. Idaho peeled and trimmed at 3-1/2 cents/ft.

April 1899

2nd Start for Lewiston by private rig this morning about 7:30 A.M. Ben Steele has 3 teams and will figure on the pole contract for about 14 miles across the mountain. About 3 feet of snow. There are about 10 miles of not bad timber. Bad piece of road at Fountains . . . This road they say is about 6 miles shorter than the old stage road via Waha and it will not be a bad road to build a line on. See Smith at Lewiston who says he can furnish us poles at Lewiston 50 ft. for $4.00 ea, 35 ft for $2.00 ea and 25 ft. for $1.25 ea. Steele thinks he can distribute them all the way out from Lewiston and Lapwai to Mt. Idaho for about 40 cents ea. **3rd** Take train from Lewiston to Pullman and see Klink. Go to Tekoa on train this P.M. and start Sandburg working on the Tekoa exchange. Pullman needs new cable 110 ft. to corner of Burgan and Jordan store. **11th** At Spokane all day. Brown and Corcoran here on underground work look-over. Favor an office for the north side. Plan to lay underground on Mill [now Wall] from Sprague to Front [now Spokane Falls Boulevard] – on Front from Mill to Division – on Division from Front to Ferry – on Sprague from Monroe to Pine – on Howard from Sprague to 4th – on Monroe

Washington north of Sprague Avenue, August 12, 1904

from Sprague to 2nd. **13th** Take Anderson and go to Davenport on train. Arrive about noon. Get team and drive to Moscow, Edwall and Sprague. Arrive at Sprague about 9 P.M. The route from Moscow to Edwall is flat along creek part of way and the whole 9 miles is hard digging. Some of it is fenced. Mr.__ at Edwall says there will be no trouble about the right of way as the farmers are all anxious to have the line built. The route from Sprague to Ritzville is hard. It will be blasting nearly all the way. **14th** Drive from Sprague to Ritzville this morning. Arrive about noon . . . think better follow the county road going south of the lakes. Do not think RR is much shorter and some of it is fenced and there is no road to team over along RR. Some say the RR is about 2 miles shorter and some say they are about the same. We will go by Harriston and Iona about a mile to the south . . . We will want about 20 50-ft. poles for the tel. office – most at Ritzville. They will take the line up to the hotel opposite the bank. Think better run in about 20 more 35s and 12 or 15 25s for exchange work. The town looks as though it should support a little exchange. Take freight train about 2:00 P.M. for Spokane. Leave Anderson to start staking back. **20th** Take train for Meyers this morning. Right of way matter fixed up. Mayor at Kettle Falls stopped men . . . go over and see him. Get fixed up promise to apply for franchise. Get team and drive over to Meyers and return to Kettle Falls for the night. Take dinner and supper at camp. **21st** Go to Meyers . . . looking for poles. The pole contractor did not do a very good job – some are small and some are rotten and crooked. **22nd** Start man out to inspect poles between Kettle Falls and Republic. Get out 2 coils #6 iron wire of postal stock for Ralph guying line to Rossland. Go to Northport by train this P.M. Kringer has nearly finished postal line. Runs out of wire about a mile from boundary. **23rd** At Northport supervising 21-man crew. Salaries range from $35.00 to $45.00/mo. **28th** Go to Juliaetta by train. Meet Casey with team and drive with him to Lewiston via Spalding and Lapwai. Looking over route for line from Juliaetta to Lapwai. Will need long poles for crossing Clearwater River at Potlatch – about 1,000 ft span. **29th** Drive to Asotin and Anatone with Casey looking up route for line Asotin to Anatone. For 2 or 3 miles out of Anatone the road winds back and forth but we can make it straight. Line will run into some hardpan on top of hill but a good share of the line will be along a good straight road. Can make a little cut off going through small town about 4 miles from Anatone. Anatone has 2 stores, blacksmith shop, stable, school house and a sort of hotel.

May 1899

2nd For 13 miles out of Baker City [Oregon] to top of summit above Auburn will have a good deal of rock work . . ., hard pan from there to Sumpter . . . some good digging but good deal of timber to cut. From Sumpter to Bourne all rock. Should have about 10 35-ft. poles at Bourne and about 25 at Sumpter. Should have a local and L. D. [long distance] cross arm on poles in Sumpter. **3rd** Leave Sumpter at 7:30 A.M. on the logging train for Baker and arrive about noon. Culled 25 25-ft. poles as rotten and 7 as too small out of pile on hand at Baker. S.V.R.R. will give us a rate of $1.50/ton on poles from Baker to McEwin and Sumpter. If can get 50 on car will cost 30 cents ea. + loading at Baker 5 cents and unloading 2 cents. Close contract with Fisk and Deardoff for dist. poles Baker to Sumpter @ 60 cents each . . . They will include 5 or 6 35-ft. poles along road. They will drop them along line near as they can and furnish horse with pole setters to snake in. **8th** Besgrove and I start for Cascade this morning with saddle horses. Raining hard. Stop about 3:00 P.M. at the Kettle River on account of rain. **9th** Start from Kettle River. Go through to Cascade. Return to Rockcut for dinner and ride to Bossburg. Arrive at 5:30 P.M. Stop at camp for supper. Road will not be too bad to build over. **10th** At Bossburg this morning trying to make settlement with pole contractor. They were to get out 765 poles @ $1.00, 25-ft long, not less than 7" across at the top and 9 35-ft poles at $2.00. Pettingill inspected the poles and reported finding 766 poles – 14 were between 6-1/2 and 7" at the top, 83 between 6 and 6 1/2" and 38 culled as too small. The contractors claim Pettingill told them the poles were too large when he inspected the final lot and blame him for the trouble. The company offered to pay for 766 poles, 9 of which are at $2.00 ea. Take train at 2:15 P.M. for Spokane. Meet Corcoran at Meyers and we arrive at Spokane at 6:30. **12th** Go to Meyers to hire Keller to take reels of wire etc. from Meyers out to camp. Take 3 red saddle horses ($24.00) for self, Holand and Long, and go out to Bryant camp. Stop tonight. **13th** Start from Bryant camp at 9:00 A.M. by horseback with . . . Arrive Republic about 5:00 P.M. Calhoun and team wants $3.50 /day. Tents for sale: 10x12 – $15, 18x45 – $80, 25x50 – $115 [other sizes and sources were also listed]. **14th** Hire cook and __ . They back out. Teams are worth $8.00 per day. Cooks want $60.00 per mo. Get prices for camp provisions and cooking utensils. While helping put out a fire in Mr. Cutle's house (in Republic) this evening get struck on the nose with a tin can. Have seven stitches put in to close up the cut. Bled a good deal and some sore. **15th** Get cook @ $50.00 per month. Buy provisions, cooking utensils and tents. Hire team to take outfit out and set up camp this evening. Hire team of Randall for $5.00 per day. We board man. **16th** At Republic. Send team out to meet pack train and get tools and material. He goes out about 10 miles and gets back with load about 7 P.M. Have 4 men digging holes and one framing poles. Dig 19 and frame 14. Raining. Make bargain for beef @ 9-1/2 cents per # for straight side. **18th** At Republic. Get check cashed and pay the bank $100.00 borrowed from them a couple of days ago. Go out to camp with pole gang this forenoon setting poles and staking line. Rain and hail. **30th** Arrive at La Grande [Oregon] about 10 A.M. Leave for Elgin and arrive there at 1 P.M. Have dinner and get team. Arrive back at La Grande at 6 P.M. [The next day. Elsom wrote extensive notes about staking out for a pole line and for a telephone office at Elgin, followed by four days of construction progress.]

June 1899
5th At Spokane all day . . . Light Co. men pull an alternator wire up against telephone cable at Maple and Pacific. Burn hole in it and disable a large number of telephones last night. **9th** Going over ground with Corcoran signing up work necessary to be done in connection with underground work and moving office. **19th** Go from Loon Lake to Colville. Put on 9 exchange arms from office north on Main St. Put in 2 pins from #9 iron L.D. loop. Put on 4 knobs for locals and wire from rest of knobs so they can be put out as needed. **24th** Arrive Arlington [Oregon] about 3:15 A.M. Get saddle horse and ride out 4 miles and count poles . . . Toby Bros P.O. Store good place for office. Wade Bros. building large new store 3/4 mile from main line . . .

Colfax telephone office in 1896. Mr. Angel is the manager.

Drive to Mayville for supper. Change teams and arrive at Fossil [Oregon] about 8 P.M. Take some snap shots on the road. Some of the road is rocky. Fossil is a nice little town. Has an opposition Exchange . . . Large store on corner opposite big hotel would be good location for office if could arrange for them to take care of it. A.B. Lamb runs drug store. J. Smith is Mayor. There is a line out to sawmill about 7 miles. Another to a ranch out about 3 miles. **29th** Arrive Endicott about 3:30 A.M. Get team and go to Colfax. Arrive there about 9:00 A.M. drive over the other road from Colfax to Mockonema, coming in at south end of Colfax. Look up old road that is now ploughed. Would be some shorter going through [this way].

July 1899

3rd Recd. from Pittsburg Reduction Co. 224 coils aluminum wire. Shipped to Portland 93 coils. To The Dalles 71 coils. To Arlington 60 coils. [Surprising that aluminum was used that early.] **14th** At Spokane all day. Look through Hyde, Jamieson, Holland and Peyton Blocks with view to putting in cable underground. Could run cable from Mill St. through the basements of Hyde Block up to ventilators haft to the roof. Carry cable through party wall through – of Traders Block to elevator shaft. Also run piece of cable along fire wall to rear of building and put up box to find Cushing and other Blocks in rear of Hyde. Run cable through Jamieson basement, taking in Jamieson elevator shaft and extending it to Holland, VanValkenburg and Peyton Blocks. Run 5 or 10 pr. cable up each shaft and multiple at each floor. Also terminate other cables on each floor. **15th** [This entry was regarding instructions for using a tripod to raise poles and set them into the ground.] Spread out on ground ready to raise by pulling on the fall of the blocks. The single leg should come toward the others and the center of tripod raised up. Should start by raising the center off the ground by hand a few feet. Make legs of timbers 6"x6" 30 or 35 ft. long. Fasten these ends by driving bar in ground and tie small hand line from foot of tripod to bar. **16th** Go to Julliaetta

Raising poles in Spokane.

and get team. Drive over route of new line from Juliaetta to Lapwai with Klink. **17th** Get team and drive from Lewiston to Anatone and return with Klink. Starting on new line. **18th** Return from Lewiston by train to Spokane office. Work this P.M. **29th** Start from Colfax for Pullman with team at 7 A.M. looking over line. Old line crosses road several times and the poles are short and shaky. Pole now broken off and hanging on wires. Should move 35-ft. poles from west side of track through town. Take north side of road and keep clear of trees . . . About 3/4 mile south move line (about 12 poles) to opposite side of road and clear trees in front of house . . . Braced poles between Colfax and Pullman – besides many of these are guyed. Pullman switchboard should have more cords . . . At times they cannot answer the calls but have to wait until cords clear out. **30th** Take train from Pullman to Uniontown and on to Lewiston . . . Need 5 or 6 50-ft. poles at Uniontown to clear trees. Line from Uniontown to Lewiston mostly in good shape. There are a number of braced poles that should be pulled back and have sills put on to hold them from pulling out. Should have a car load of 50-ft. poles distributed between Spokane and Garfield as follows: 10 at Spangle, 5 at Plaza, 6 at Rosalia, 12 at Oakesdale, 12 at Belmont and 15 at Garfield. **31st** Hire team and go out to Klink's camp this morning. Stay till after supper. The line is complete to stake 71. Poles are set to 78. All holes are in rock. Has 11 hands. Drive back to Lewiston tonight. Arrive about 9 P.M.

August 1899

11th At Spokane all day. Besgrove came in. **12th** Go to Pomeroy. Send 4 linemen to WW to start T.C. 50. Stay at Pomeroy tonight. Find about 50 old 25-ft. poles at Pomeroy to raise line over trees. Cable pole should be stubbed and L.D. wires run into office for testing. **13th** At Dayton. Needs new piece 20 pair cable from "x con frame" to pole 125 ft. The switchboard is like Heppner. Has 3 rows of drops . . . 2 rows of jacks, 20 in row #1 to 40. Need about 20 60-ft. poles on 1st and 2nd St. Trees are bad both sides of Main. There are 35-ft poles now . . . Leave Dayton on NP at 6:15 P.M. Go to WW and arrive about 8:30. Take 10:45 P.M. train for Spokane.

The northeast corner of Spokane's First Avenue and Howard Street, circa 1904.

September 1899
9th Supervisor [Mr. Corcoran] from Portland at house for dinner.

The whereabouts of diaries covering 1900 through 1914 are unknown except for the period from May 1911 to December 1912, when he was starting to farm at Saltese. On January 5, 1911, he quit the phone company but returned on June 12, 1913. These were troublesome times for him. He was very disappointed in being replaced as district superintendent with a pay-cut from $175 to $150 per month. Although a quiet person who kept things to himself, he must have suffered during this period under new management. His diary entries then resumed, but with only a relatively few entries about his activities as title examiner and telephone right-of-way agent.

1915 Diary Entries
December
18th Division Street bridge fell. Five men killed and eight or ten injured. [He took pictures the next day.] **21st** Attended conference in Seattle on history of telephone toll lines in Washington.

Collapse of the Division Street Bridge, December 18, 1915.

This sign, erected in 1910 at Second and Stevens, was announcing the construction of the seven-story Pacific Telephone and Telegraph Company building to be constructed on this site, which became its headquarters.

Telephone building at Third and Crestline, February 26, 1911.

It was 1915 before the distance between New York and San Francisco was spanned by telephone lines. On this memorable occasion Bell and Watson repeated their conversation of March 10, 1876: "Mr. Watson, come here, I want you."

From 1905 to 1915, there had been two major telephones companies serving Spokane, but in 1915 the Pacific Company acquired the property of the Home Telephone and Telegraph Company. It operated under the name of the Home Company until 1935 when the name was changed to Pacific Telephone and Telegraph Company. By 1916 Elsom himself was involved in transcontinental telephoning.

1916 Diary Entry

January
10th Rehearsed a demonstration of talking to New York from Spokane by phone. Could hear N.Y clearly. [Elsom spent much of this year and through 1923 negotiating for phone right-of-ways.]

Elsom's diary entries for this period are brief and not as revealing as the earlier diaries. However, items by and about Elsom appearing in the telephone journal *Washington Translator* provide information on his life and work for the telephone company during this period. In 1921 he attended a national convention of the Telephone Pioneers in St. Louis. The following letter dated November 8, 1921, and subsequently published in *Washington Translator,* give an account of this trip, which obviously had both professional and personal significance for him.

> Mr. H.J. Tinkham,
> Division Superintendent of Plant,
> Seattle, Wash.
>
> Dear Sir:
>
> I have just returned from the St. Louis trip and had a fine time. The Southwestern Co. took great pains to entertain us, furnished automobiles for the entire crowd for an excursion about the city, etc. The Western Elec. Co. provided us with a fine dinner out at Riverside Club.

The right-of-way force in 1924. From left, Mr. Winget, Thomas Elsom and Mr. Dunton.

The A.T. and T. Co. gave us a demonstration of the Loud Speaker, which is surely a wonderful thing, and they also gave us a swell banquet at the end of the meet. Mr. Thayer was there and presided. Mr. Carty was elected next president, the next convention to be at Cleveland.

After the close of the convention, I sort of played "hookey" for a few days and swung around to the old boyhood home in New York state and took a kodak picture of the old swimming hole, little red school house, etc., and also made a call at 195 Broadway, New York City, the home of the A.T. & T. Co. on the roof of which I took a kodak picture of the statute on the roof, and as it happened to be the morning on which Marshal Foch arrived in this country, they were raising a big new flag on the new flag pole and I took a snap of the flag as it was being raised. A new addition is being made to the building and the new pole in on the new part of the building and men were still working on the pole. This was the first time a flag had been raised on this pole.

. . . In going through the building, I passed a door labeled A.H. Girswold, Assistant Vice-President. I stepped into the office but the colonel was out. I left a card and since returning have received a nice note from him.

I went to Chicago, presented Mr. Fullerton's letters to Mr. Kennedy and Mr. Redmund and was very kindly received by each of these gentlemen. Spent considerable time at the Hawthorn plant, where the Western Elec. Co. provided me with a guide who took pains to show and explain things to me. It is certainly a great institution and has to be seen to be appreciated, for no amount of reading matter can furnish any appreciation of its immenseness.

I saw the big installation of machine switching that is being placed for the Chicago Co. Mr. Redmund says he expects it will be about a year before it is ready to cut in.

Two of the delegates from the Pacific Co., Mr. Klink and Mr. Welsh are old-time plant men of Spokane, whom I was glad to meet. The rest of the Pacific crowd were very pleasant people, especially Miss Blythe, of the San Francisco office, who took great interest in getting our crowd together for a picture, which she is taking home for the magazine. It was decided that Mr. Ogan, of Los Angeles, should write the story for the magazine.

I spent about two days at my old home, father and mother are getting old, father 81 years, and not very strong. They were greatly pleased. All together it was a fine trip which I am very grateful for and have written Mr. McFarland and Mr. Fullerton to that effect. Thank you kindly, I am,

Yours truly, (signed) Thos. H. Elsom

1924 Diary Entries

January
7th Picture of C.B. Hopkins framed and hung in the company assembly hall.

July
26th Nice party at Davenport Hotel in honor of my 35th year with the telephone company.

November
5th Coolidge elected President.

1925 Diary Entry

February
12th Out on line between Elberton and Garfield tonight.

1926 Diary Entries

March
10th Today is 50th anniversary of telephone. First conversation was March 10, 1876.

May
4th Father died this morning.

1927 Diary Entry

June
20th Electric rail service stopped to Liberty Lake. [This was a major impact on him because he had used it to commute about fifteen miles to the Spokane office for many years. He walked to the Greenacres stop over a mile from his Saltese home in spite of weather extremes. He waved off offers of rides by his neighbors, saying that he wanted the exercise.]

Electric train station at Greenacres, where Elsom commuted to Spokane. (Photo EWSHS, L94-19.32.)

Thomas Elsom (right), right-of-way supervisor, with Katherine Wall, past chief operator, and Burt Callison, district manager, in 1928.

Chapter 4
Retirement and Family Photo Album

On September 1, 1930, after nearly 44 years of service, Thomas Elsom retired from the telephone company with a pension of $87.00 per month. During his years with the telephone company, he had served in a number of capacities. When the Inland Company had taken over in 1890, he became foreman of construction. In June 1891, the Pacific Company took over and he became chief foreman. He had the title of superintendent of construction in 1898 and supervised telephone system expansion over the entire Northwest as recorded in earlier chapters. He became division construction foreman in 1905 and district superintendent in 1910.

On January 5, 1911, he had quit the company for two years because of his great disappointment in being replaced as district superintendent, but returned to the company two years later. The last fifteen years, he was in the engineering department as a right-of-way agent.

The October 1930, issue of *Washington Translator* **contains two items relating to Elsom's retirement. The first is a letter from Elsom dated August 29, 1930:**

> Mr. C.D. Phillips,
> District Plant Manager,
> Spokane, Washington.
>
> Dear Mr. Phillips:
>
> On retiring from active service with the Telephone Company I want to express to you and through you, to Mr. Larsen, my appreciation of the courteous consideration accorded me in our relations with each other in the service of the Telephone Company.
>
> It lacks but a little of being forty-four years since I entered the employ of the Company. In looking over the trail it does not seem very long or too rough, but mostly pleasant traveling, particularly the recent years.

Thomas and Nell at Saltese ranch in 1920.

Elsoms' ranch home on 80 acres at Saltese where they lived from 1911 to 1932. Photo taken in 1919.

It seems to me that the Telephone Company is a pretty good institution to tie up with for a person who is disposed to keep his traces tight and do the square thing.

My regrets are that conditions do not permit of further continuance in the game. In breaking away I sincerely wish you each the best possible success.

Sincerely yours, (signed) Thos. H. Elsom

The second appears in a column relating to the activities of various employees. It is a rather low-keyed reference to his departure from his office upon retirement.

T.H. Elsom has left us, slipping away in his typical fashion, leaving only a note on the bulletin board wishing us good-bye and good luck. Of course we are sorry to see him go but we are proud of his unclouded record. After his years of unselfish service it is only just that he should rest. He takes with him in his retirement the esteem and respect, not only of every man in the Engineering Department, but of every man and woman in the telephone business. We, his immediate fellow workers, wish him all happiness and good luck and each one of us holds the hope that when we total the last estimate and retire from active service, we may look back upon as honorable a record as Thomas H. Elsom.

In September 1932, after Thomas's retirement, they moved from their home at Saltese to a one-acre place in Greenacres at 17602 East Indiana, which he farmed as a hobby. He also participated in church and civic affairs. Nell and Thomas celebrated their 60th wedding anniversary there. When this place became too much for them to handle in their advanced years, they moved into Spokane. Their final years were at 2943 North Lacey Street, two houses away from their daughter, Gertrude, who could readily help care for them.

A Few Remaining Diary Entries of Significance

July 10, 1933 Received telegram from Wilson [brother] that mother died this morning.

Nell Pratt (Elsom) in 1888.

The Elsom family on Nell and Thomas's 60th wedding anniversary on October 17, 1951. Front row, from left: Gertrude (Elsom) Ladd, Nell Elsom, Joan Elsom (Burnard's daughter), Thomas Elsom and Russell Elsom. Back row: Ruth (Elsom) Ainsworth, Floral (Elsom) Stephenson and Burnard Elsom.

April 21, 1934 Go to IOOF meeting in evening. Lodge presents me with a 30 year badge.

January 6, 1936 Burnard [son] goes to work for telephone company in telegraph/teletype dept.

October 3, 1949 Check up on lot at Greenwood Cemetery. Space for eight graves – three occupied.

January 10, 1950
I decline to run for Township Supervisor. Larson is elected to succeed me as Supervisor.

Thomas Elsom's final diary entry (Shoveling paths. Cold and clear) was on December 31 1951. He died two years later on December 15, 1953 at age 88 from pneumonia following complications with a stroke and a broken hip. Nell had preceded him in death a few months earlier at age 84. They were buried at Greenwood Cemetery.

His son Burnard, who later also retired from a long career with the phone company, wrote in his diary:

> It's a rainy dark day. Dad's funeral was at 1 P.M. The Odd Fellows took part in the ceremony at the funeral home and the Masons at the cemetery. It seems right to have dad and mother together again. Although this was a dark day, there are some bright spots. The funeral was fitting and we all feel that the memories we have are not to be taken from us. We have spent several hours talking over old times.

Thomas Elsom was a Past Master of Spokane Masonic Lodge #34 in 1908. He was also a member of the Wilford Chapter of the Odd Fellows Lodge, the Electra Chapter of Eastern Star, the National Grange, the Telephone Pioneers of America and the Euclid Avenue Baptist Church. He had 12 grandchildren and 14 great grandchildren.

Thomas and Nell Elsom on their 60th wedding anniversary in 1951.

Crossing The Bar

Sunset and evening star,
 And one clear call for me!
And may there be
 no moaning of the Bar
  When I put out to sea.

* * * * *

For tho' from out our bourne
 of Time and Place
The flood may bear me far,
I hope to see my Pilot
  face to face,
When I have crost the bar.

— Tennyson

In Memory of

Thomas H. Elsom

Born August 11, 1865

Passed Away December 15, 1953

Services

Sunset Memorial Chapel

Spokane Washington

December 19 1953 1:00 P. M.

Reverend Ransom D. Marvin

Officiating

Soloist Phil Crosbie

— — — —

Interment

Greenwood Cemetery

Spokane Washington

Elsom family on a picnic on Mica Mountain on August 25, 1912. From left, Nell holding Burnard, Ruth (front), Floral, Gertrude and Russell.

The Elsom children in 1941. From left, Ruth Ainsworth, Russell, Gertrude Ladd, Burnard and Floral Stephenson. Burnard later also retired from long service with the telephone company.

At the Elsom home at Saltese on August 16, 1920. From left (standing), Gertrude Elsom Ladd, Ruth Elsom, Floral Elsom. From left (seated), Myron Ladd, Burnard Elsom and Nell Elsom.

Thomas and Nell Elsom with their first grandchildren, Dean Ladd, Floral Ann Elsom and George Ladd, in 1924

Elsoms' Saltese home, as improved in 1927.

Thomas Elsom's parents, Joseph and Jane Elsom, circa 1925.

Family picnic at Liberty Lake in 1902 when Thomas's mother and father were visiting from South Dakota. From left, Thomas's mother, Jane Elsom; his son Russell and daughter Gertrude; Thomas (sitting); and Thomas's father, Joseph Elsom.

Thomas and Nell Elsom's last home, located at 2943 North Lacey.

A T & T Co. 109
Ainsworth, Ruth J. (Elsom) 61, 62, 115, 118, 119
Almota, Wash. 16, 17, 20, 74, 75
American Bell Telephone Co. 19, 21, 24
Anatone, Wash. 98, 99, 103
Arlington, Oregon 76, 90, 100, 101
Asotin, Wash. 85, 98
Athens, Oregon 77
Auburn, Oregon 99

Baker City, Oregon 74, 99 Bank of, 49
Bean, Walker L. 37
Becker Buick 28
Bell Telephone Co. in the Pacific State 16, 27, 56
Bell, Alexander Graham, 11-13, 15, 26, 32, 107
Berliner Transmitter 16, 37
Big Bend Country 21, 77, 85
Black Mariah 46
Blake Transmitter 15
Blake, Francis Jr. 14
Blue Creek Canyon 49
Blythe, Miss 109
Bossburg, Wash. 99
Bourne, Oregon 99
Brewster, Wash. 49, 83

Bridgeport, Wash. 84
British Columbia 21, 83
Brooklyn, NY 26
Browne, J.J. 27, 97
Browne's Addition 40
Bryant Camp 99, 100
Bunker Hill Mine 26
Burgan & Jordan Store 97
Burke, Idaho 26, 47, 48, 51

C & C Flouring Mill 26
Callison, Burt 111
Canadian Border 34
Canadian Pacific 23
Cannon, A.M. 24, 27
Carlin, Col. 25
Carlton, Orleans County NY, 32
Carty, Mr. 109
Cascade Mountains 34, 59
Cascade, Oregon 99
Cataldo Mission 26
Cataldo, Idaho 48
Centennial Exposition in Philadelphia 11
Chelan Falls 83
Chelan, Wash. 79, 83
Cheltenham, Gloustershire, England 21, 24
Cheney, Wash. 24, 82
Chicago & Northwestern Railway 37
Chicago, Illinois 109
Chinamen 61

Civil Engineering 36
Cle Elum, Wash. 79, 81
Cleveland, Ohio 109
Coeur d'Alene, Idaho 20, 25, 26, 37, 44, 45
Coleman, Mrs. 44
Colfax Flood 94, 95
Colfax Office 101
Colfax, Wash. 16, 17, 20, 21, 24, 74, 85, 93, 101, 103
Colville, Wash. 100
Conconully, Wash. 83
Cook, Francis 36, 78
Coolidge, Pres. 110
Corbin, D.C. 23-26, 47
Corcoran, C.J. 59, 92, 97, 99, 100, 105
Cottonwood, Idaho 96
Coulee City, Wash. 78, 79, 81
Coxey's Army 28
Crawford, George 36
Crescent Block 55
Creston, Wash. 78
Cushing Building 55,
Custer Massacre 11
Cutle, Mr. 100

Davenport Hotel 27, 110
Davenport, Wash. 20, 77, 78, 86, 98
Dayton, Wash. 74, 75, 103
Deardorff 99
Deep Creek, Wash. 77, 81

123

Dennis, G.B., 27
Division Street
  Bridge 105
Dodd Block 63
Donald & Davis 83
Downs, Wash. 86
Dunton, Mr. 108
Durham, N.W. 25
Dyer, Wash. 84
Eagle Block 55
East Helena, Mont. 26
Eastern Star, Electra
  Chapter 116
Edison Electrical
  Illuminating Co.
  25, 26
Edwall, Wash. 98
Egypt, Wash. 77, 78
Elberton, Wash. 110
Electric Bond &
  Share Co. 25
Elgin, Oregon 100
Ellensburg, Wash. 32
Elsom Diaries 34-38,
  40, 42-44, 49, 51,
  53, 56, 90
Elsom, Burnard H.
  61, 115, 116, 118,
  119
Elsom, Floral Ann
  120
Elsom, Floral H. (see
  Stephenson, Floral)
Elsom, Gertrude M.
  (see Ladd, Gertrude)
Elsom, Joan 115
Elsom, Joseph & Jane
  (Harmer) 32, 33,
  121,122
Elsom, Joseph

Lawton 61-63
Elsom, Nell (Pratt)
  59, 60, 61, 62,
  113-120, 122
Elsom, Russell L. 61,
  62, 63, 115, 118,
  122
Elsom, Ruth J. (see
  Ainsworth, Ruth)
Elsom, Thomas 21,
  22, 28, 29, 31, 32,
  34, 36-40, 44, 45,
  56, 59, 61, 62, 63,
  67, 73, 98-105,
  107-109, 111, 113-
  116, 120, 121, 122
Elsom, Thomas Jr.
  61, 63
Elsom, Wilson 114
Elsoms' Homes 62,
  63, 113, 120, 122
Emperor of Brazil 11
Endicott, Wash. 101
Ephrata, Wash. 86
Espinola, Wash. 86
Euclid Avenue
  Baptist Church
  116
Eureka, Wash. 77
Exposition Building
  28
Falls City Opera
  House 55
Farmers Telephone &
  Telegraph Co. 81
Ferginson Camp 86
First National Bank
  Building 26, 39,
  45, 54
Fisk 99

Foch, Marshal
  (French General)
  109
Fort George Wright
  28
Fossil, Oregon 101
Fountain family 96
Fourth of July
  Canyon, 26, 56
Frankfurt Block 55
French Phone 15
Frisbie, Miss 51,53
Fullerton, Mr. 109
Furth, Mayor Fred 55
Garfield, Wash. 85,
  91, 93, 103
General Electric
  Industry of
  America,
  (General Electric
  Co.), 25
Genesee, Idaho 85
Glover, James, 27
Gomer, Mr., 35
Grand Hotel 55
Grangeville, Idaho
  93,96
Granite Block 55
Gray, Elisha 14
Great Northern
  Railroad 85, 86
Great Spirit 49
Greenacres, Wash.
  110, 111
Greenwood Cemetery
  116
Griswold, A.H. 109
Harrington, Wash.
  74, 85-6
Hartline, Wash. 78,

81, 86
Hatley, 93
Hauser Junction, Idaho 25, 37
Hazel Block 56, 91
Henry, Joseph 11
Heppner, Oregon 103
Highmore, Dakota Terr. 35
Hitchcock, Dakota Terr. 35
Hodge, Miss 68
Holland Block 101,102
Holmes School 63
Home Telephone & Telegraph Co. (Home Co.) 107
Hopkins, C.B. 16-21, 24, 27, 29, 36, 38, 45, 56, 59, 85, 90, 110
Hurley, Dakota Terr. 35
Huron, Mont. 35, 37, 38, 44
Hyde Block, 55, 101, 102

Idaho (state) 17, 28, 24, 36, 49, 59, 90
Idaho State University 85
Inland Telephone & Telegraph Co. 21, 22, 27, 32, 40, 45, 56, 59, 90, 112
Inland Territory 74
Isaacs, Mr. 77

Jacoy, Peter 63
Jamieson Block 101, 102
John Day, Oregon 74
Jorgenson, Dr. Joseph, 7
Juliaetta, Idaho 85, 98, 102, 103

Kellogg, Idaho 24
Kellogg, Lucien E. 16
Kelvin, Lord 11
Kendrick, Idaho 85
Kennedy, Mr. 109
Kettle Falls, Wash. 91, 92, 98
Kingston, Idaho 45
Klink, Mr. 93, 97, 103, 109
Kootenai (a steamer) 23
Kringer, Mr. 92, 93, 98

La Grande, Ore. 90, 100
Ladd, Dean 120
Ladd, George 120
Ladd, Gertrude M. (Elsom) 20, 59, 61, 62, 63, 73, 90, 114, 115, 118, 119, 122
Ladd, Myron 119
Lake Chelan, Wash. 83
Lake Coeur d'Alene, Idaho 26
Lamb, A.B. 101
Lamona, Wash. 85, 86
Lapwai, Idaho 97, 98,103

Larson (township supervisor) 116
Larson, Mr. 112
Latah, Wash. 91
Lewiston, Idaho 19, 20, 84, 85, 92, 93, 96-98, 103
Liberty Lake, Wash. 110
Lind, Wash. 82
Little Big Horn 11
Little Dalles, Oregon 23
Loon Lake, Wash. 100
Los Angeles, Calif. 109

Masonic Lodge #34 116
Mayview, Wash. 75
Mayville, Oregon 101
McCartney, H.H. & Co., 23
McEwin, Oregon 99
McFarland, Mr. 109
Medical Lake, Wash. 82
Meyers (Falls), Wash. 91, 92, 98
Mileage to various RR Stations 92
Military Telephone & Telegraph Line 26, 45
Mining Camps 22, 26, 38, 45, 51, 59, 61
Minnesota 34
Missoula, Mont. 25, 45
Mitchell, S.Z. 25

Mockonema, Wash. 101
Mondovi, Wash. 77
Monler, M. 64
Montana (state) 21, 24, 34, 35, 37
Montana Syndicate 25
Moore, Paul F. 24
Moran Prairie 23
Moscow, Idaho 85, 98
Mt. Idaho 96, 97
Mt. Missoula 16, 20
Mullan Road 25, 45
Mullan, Idaho 26, 47, 53
Munch Treatment, Dr. 92
Murray, Idaho 26, 47, 51
Nash, Lucius G. 40
Natatorium Park 63
National Grange 116
New York State 109
New York, NY 107
Newberry, A.A. 23, 25
Nez Perce War 16
Nichols Block 25
Nichols, Burt 25
Norman W.S. Telephone System 24
Norman, Ben, 27
Norman, William S. "Billy" 21, 22, 24-29, 45, 49, 50, 53, 56
North Yakima, Wash. 79
Northern Pacific Railroad 16, 24, 37, 38, 45, 55, 59, 82, 83, 85
Northport, Wash. 23, 98
Northville, Mont. 32, 35, 44
Northwest Magazine 47
Noxon, Mont. 44
Oak Orchard, NY 32
Oakesdale, Wash. 85, 93, 103
Odessa, Wash. 85 86
Ogan, Mr. 109
Okanogan, Wash. 79, 83
Old Mission, Idaho 49, 50
Old National Bank 61
Olympia, Wash. 32
Oppenheimer, Francis 23
Oppenheimer, Joseph, 23
Oregon (state) 17, 23, 34, 73, 77, 90
Orondo, Wash. 79
Osborne, Idaho 51
Pacific Co., 107, 109, 112, 114
Pacific Coast 47
Pacific Hotel 155
Pacific Northwest 40
Pacific Northwesterner, The (Vol. 33 #2) 21
Pacific Telephone & Telegraph 107
Pacific Telephone Magazine 37, 59
Palouse Gazette 16, 17, 19
Palouse, Wash. (city and area) 20, 85, 91
Pasco, Wash. 82, 83
Pateros, Wash. 79, 83
Peacefull Valley, Wash. 63
Peltier, Jerome 21, 23, 27
Pendleton, Oregon 77, 90, 93
Peshastin, Wash. 80
Pettingill, Mr. 99
Peyton Block 101, 102
Philistines, Land of 36
Phillips, C.D. 112
Pittsburg Reduction Co. 101
Plaza, Wash. 103
Pomeroy, Wash. 74, 90, 93, 103
Port Townsend, Wash. 32
Portland, Oregon 16, 21, 73-75, 77, 91, 101, 105
Post Mill 26
Potlatch, Lewiston Mill 98
Pratt, Nellie (see Elsom, Nell)
Prescott, Wash. 77
Prosser, Wash. 82
Provision prices 92,

93
Pullman, Wash. 85, 96, 97, 103

Ralph, Mr. 93, 98
Randle & Team 100
Redmund, Mr. 109
Republic, Wash. 91, 92, 98, 100
Republican Party 28
Revelstoke, B.C. 23
Ridpath Hotel 91
Riperia, Wash. 75
Ritzville, Wash. 82, 98
Riverside Club 107
Riverside, Wash. 83
Rochester, NY 59
Rockcut, Wash. 99
Rockford, Wash. 91
Rocky Mountain Bell Co. 74
Rohde, Mr. 91,92
Rosalia, Wash. 103
Ross Park Line 27
Rossland, B.C. 98

Sacred Heart Hospital 37
Saltese, Wash. 63, 105, 110, 114
San Francisco, Calif. 17, 19, 74, 90, 107, 109
Sandburg 91, 97
Seattle, Wash. 32, 77, 79, 81-83, 105, 107
Sherman 20, 25, 26, 45
Signal Service 16
Smalley, Eugene V.

47
Smith, J. 101
Soap Lake, Wash. 86
South Dakota Territory 32-36, 44
Southwestern Co. 10
Spalding, Fred 25
Spangle, Wash. 73, 103, 81-83, 85, 90, 92, 93, 98-100, 103, 107
Spokan Times 16
Spokane & Northern 24
Spokane & Palouse Railroad 37
Spokane (also Spokane Falls), Wash. 17, 19-25, 27, 29, 34, 36, 44, 45, 47, 49, 51, 53, 56, 73-77, 79,
Spokane Cable Railway System 27
Spokane County Courthouse 24
Spokane Electric Light & Power Co. 37, 38
Spokane Exchange & Subscribers 38
Spokane Falls 1890 Industrial Exposition 28
Spokane Falls Board of Trade 53
Spokane Falls Illustrated 53
Spokane Falls Review

47, 51, 55
Spokane Great Fire 27, 53, 54, 57, 58
Spokane Hotel 27, 91
Spokane Indian 61
Spokane National Bank Bldg. 63
Spokane Streets & Buildings 29, 30, 39, 97, 102, 104
Sprague, Wash. 82, 98
St Paul, Minn. 16, 47
St. Louis, MO 107
Stampede Tunnel 81
Starbuck, Wash. 75
Steele, Ben 96
Stephenson, Floral H. (Elsom) 61, 62, 115, 118, 119
Stowger Automatic Phone & Wall Set 15
Sumpter, Oregon 99
Sunset Telephone & Telegraph Co. 27, 56

Tacoma, Wash. 32
Tekoa, Wash. 92, 97
Telephone Buildings, Exchanges & Switchboards 66-69, 76, 106
Telephone Line Crew & Tools & Equipment 65, 87-89
Telephone Pioneers 107 116
Telephone Test Stations 50

Thompson Falls, Mont. 44
Thompson's Ranch 49
Thresher, Frank L. 53
Tinkham, H.J. 107
Toby Bros. PO & Store, 100
Traders Block 102
Tull Block 55
Twisp, Wash. 79
U.S. Government 38, 51
Umatilla, Oregon 74
Union Iron Works, 83
Union Town, Wash. 85, 103
Union, Oregon 90
Upper Falls Plant 26
Van Valkenburg Block 102
Wade Bros. 100
Waha, Idaho 96, 97
Wall, Katherine 111
Walla Walla, Wash. 17, 16, 19, 20, 25, 45, 73, 76, 77, 82, 93
Wallace, Idaho 24, 47, 51-53, 56
Wallace, Mr. 61,
Wallace, Mrs. 53
Wallula, Wash. 74, 82, 96
War Dept. 25
Wardner Junction, Idaho 53
Wardner, Idaho 20, 21, 26, 45-47, 49, 53
Washington Block 55
Washington State Agricultural College 85
Washington Statehood 51
Washington Translator 107, 112
Washington Water Power Co. 26, 27
Washington, DC 28
Washington, State & Area 16, 17, 20-24, 32, 36, 38, 59, 90, 105
Waterville, Wash. 78, 79, 81, 83
Watson, Thomas 11 13, 107
Waverly, Wash. 92
Welch, Mr. 109
Wenatchee Junction 81
Wenatchee, Wash. 79-81, 83
Western Electric Hawthorne Plant 109
Western Union Telegraph Co. 16, 61, 107, 109
Westlake, Idaho 96
White Truck 70
Wilbur, Wash. 78
Wilson Creek, Wash. 86
Windsor Hotel 55
Winget, Mr. 108
Wolf Lodge 49, 50
World Encyclopedia 35
Yakima, Wash. 32, 82, 83